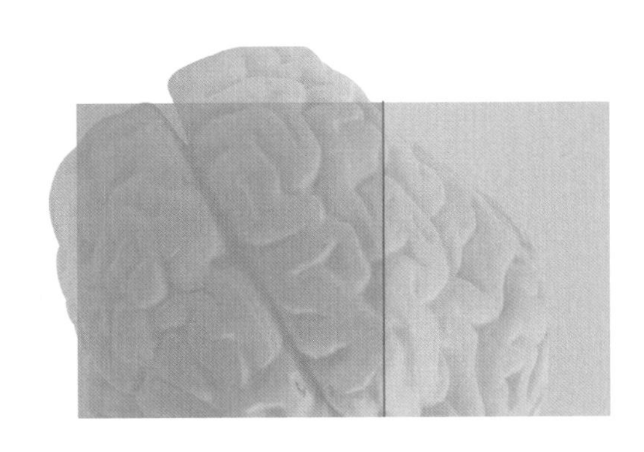

The Forebrain

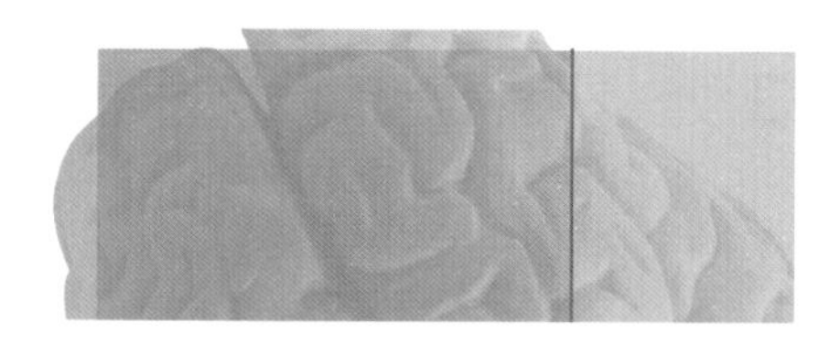

Brain Disorders

Cells of the Nervous System

Learning and Memory

Meditation and Hypnosis

The Forebrain

The Hindbrain

The Midbrain

The Neurobiology of Addiction

The Spinal Cord

Sleep and Dreaming

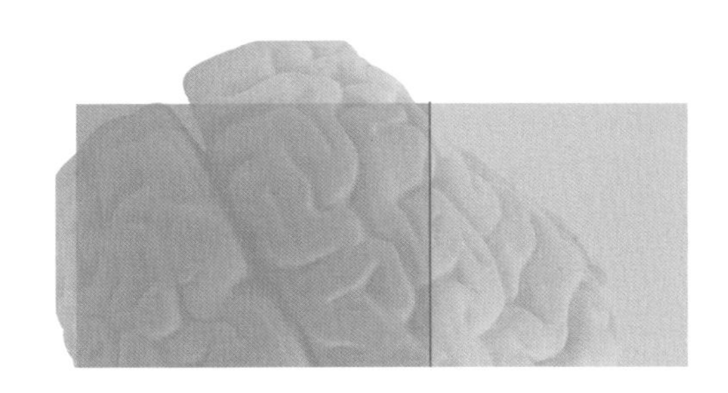

The Forebrain

Elizabeth Tully

CHELSEA HOUSE PUBLISHERS

VP, NEW PRODUCT DEVELOPMENT Sally Cheney
DIRECTOR OF PRODUCTION Kim Shinners
CREATIVE MANAGER Takeshi Takahashi
MANUFACTURING MANAGER Diann Grasse
PRODUCTION EDITOR Noelle Nardone
PHOTO EDITOR Sarah Bloom

STAFF FOR THE FOREBRAIN

PROJECT MANAGEMENT Dovetail Content Solutions
DEVELOPMENT EDITOR Carol Field
PROJECT MANAGER Pat Mrozek
PHOTO EDITOR Robin Landry
SERIES AND COVER DESIGNER Terry Mallon
LAYOUT Maryland Composition Company, Inc.

A Haights Cross Communications Company ®

www.chelseahouse.com

First Printing

10 9 8 7 6 5 4 3 2 1

Library of Congress Cataloging-in-Publication Data

Tully, Elizabeth.
The forebrain / Elizabeth Tully.
p. cm. — (Gray matter)
Includes bibliographical references and index.
ISBN 0-7910-8509-0
1. Prosencephalon. I. Title. II. Series.
QP382.F7T85 2005
612.8′25—dc22

2005011688

All links, web addresses, and Internet search terms were checked and verified to be correct at the time of publication. Because of the dynamic nature of the web, some addresses and links may have changed since publication and may no longer be valid.

Contents

1 Don't Take the Forebrain for Granted

The human brain is a marvel. This gelatinous, wrinkled mass of tissue, weighing only about 1400 grams, or 2% of a person's total body weight, is responsible for every thought, memory, and emotion we experience. To see the approximate size of your brain, place your fists together with the knuckles touching each other. Imagine this 3-pound organ floating in **cerebrospinal fluid** inside your skull. Without your brain, you could not live the life of a human being. Although many organs of the body have the potential to be transplanted, including the heart and kidneys, no one has figured out how to replace a damaged brain. Your brain gives you a unique personality. It makes you who you are.

This book will introduce you to basic neuroanatomy, a subject that is difficult for many students to understand at first. As you read these chapters, it may help you to consult a neuroanatomy atlas and to make sketches of the different parts of the brain. Over time, you will begin to master the terminology and relationships of basic neuroscience. In this book, we will focus on learning about the structure and function of the **forebrain**.

The forebrain is the top part of the brain, which forms during the early embryo stage of human life. During the first month of pregnancy, the brain develops as a series of three small swellings called the forebrain, the midbrain, and the hindbrain. The forebrain then divides into two more swellings, the **telencephalon** and the **diencephalon.** The telencephalon develops into the **cerebral hemispheres.** The cerebral hemispheres make up about 80%–90% of the brain's total mass. The **cerebral cortex**, the thin outer layer of these hemispheres, contains the connections that are responsible for the highest order of thinking (or cognition), as well as memory and language. The diencephalon develops into the **thalamus** and **hypothalamus.** The thalamus is important because it is a relay station that sends sensory information to the cerebral cortex. The thalamus helps us to focus our attention on specific sensory systems. The hypothalamus is also a very important gatekeeper that controls the body's autonomic and endocrine activity (involuntary muscle movement and hormones). As you can see, the forebrain is a large and complex part of the brain, and it plays many critical roles in our functioning (Figure 1.1).

While the field of neuroscience is advancing rapidly around the world, in part due to new technologies, many questions remain unanswered about the brain and its disorders. In this book, we will explore disorders of the forebrain such as Alzheimer's disease, **Parkinson's disease**, attention deficit hyperactivity disorder (ADHD), stroke, and schizophrenia. These disorders affect many millions of people around the world, and there are no cures available yet. These disorders are also central to our discussion because they highlight certain aspects of normal brain function. Many neuroscientists around the world are working to understand the functions of the normal brain as well as the brain's disorders. **Neurologists** are medical doctors who specialize in diagnosing and treating diseases of the brain and nervous system. **Psychiatrists** are medical doctors who specialize in treating both physical and mental disorders of the brain. Together, these researchers are discovering exciting new knowledge about the brain.

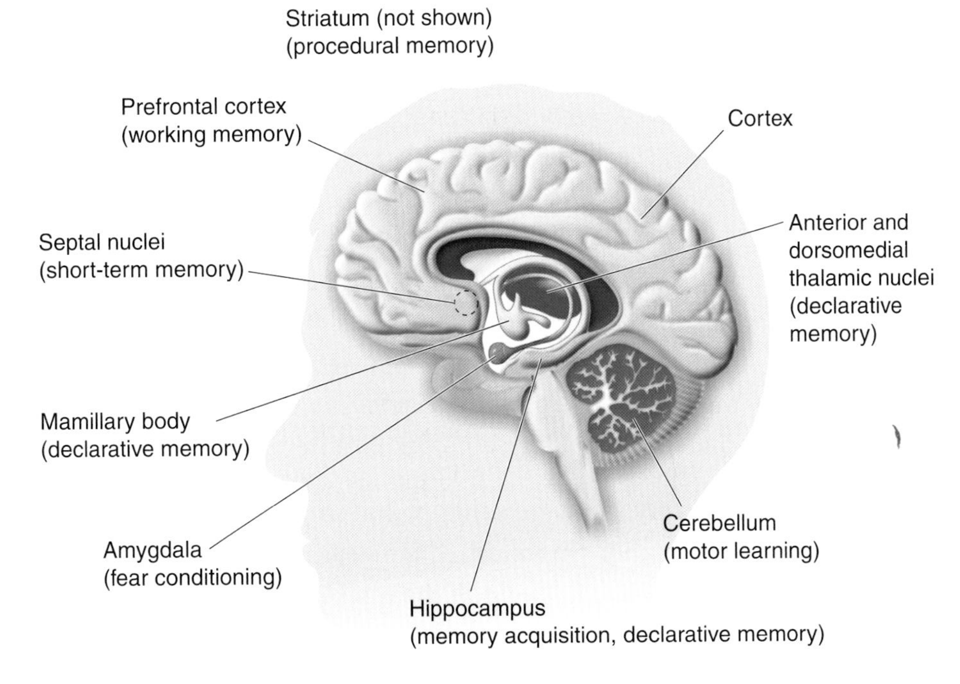

Figure 1.1 The structures of the forebrain play many critical roles, including memory and learning.

As you learn about the forebrain, you will be surprised by the complexities of an organ that you may have taken for granted. You will gain a greater appreciation for the challenges faced by people with the brain disorders listed above. You may wonder, as many neuroscientists do, what causes schizophrenia? Is Alzheimer's disease preventable? Do toxic chemicals cause ADHD? You may reflect on your own brain health and decide to protect yourself by taking such measures as wearing a helmet when you ride your bike, avoiding alcohol and street drugs, and getting vaccinated against meningitis.

■ **Learn more about disorders of the forebrain** Search the Internet for *Alzheimer's disease, Parkinson's disease, attention deficit hyperactivity disorder (ADHD), stroke,* or *schizophrenia.*

2 Dementia and Alzheimer's Disease

Are you forgetful? Do you sometimes forget where you left your sunglasses, your backpack, or your homework assignment? Everyone forgets things, and usually in a short time we can remember them again. However, for people suffering from **dementia**, a disease of the forebrain, memory lapses become serious and permanent. Try to imagine what it would be like to forget your way home, how to dress yourself, or the names of your loved ones.

Dementia is a progressive decline in mental functioning that includes severe memory impairment. The term describes a group of symptoms that are caused by many conditions. **Alzheimer's disease**, **multi-infarct dementia**, and **Lewy body type dementia** are three of the most common forms of dementia in older people.

In multi-infarct dementia, many areas of the brain die when a small series of **strokes** cuts off their blood supply. A stroke usually happens suddenly and gives rise to rapid changes in a person's functioning. One of the most common causes of a stroke is high blood pressure, or **hypertension**. Strokes can be easily prevented by treating and lowering high blood pressure. The severity of multi-infarct dementia (unlike Alzheimer's disease) may fluctuate over time.

Alzheimer's disease affects mostly the basal forebrain, an amazing part of the brain that will be described in greater detail

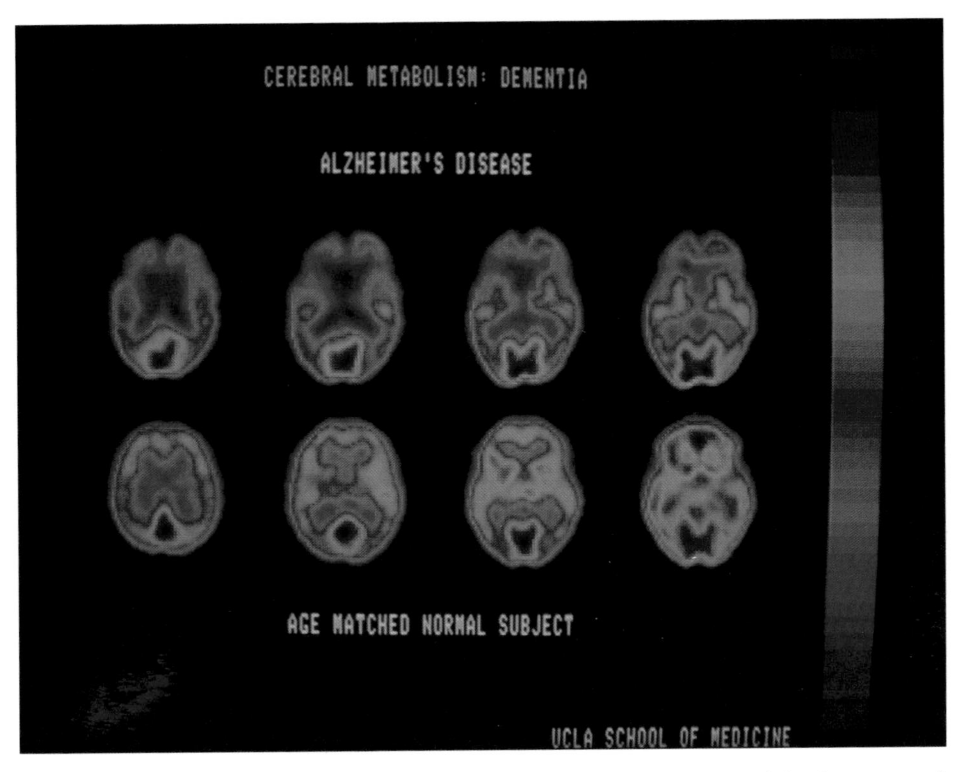

Figure 2.1 PET (positron emission tomography) scans showing activity in a normal brain and in the brain of a patient with Alzheimer's disease. Highly active areas appear red, moderately active areas appear yellow, and less active areas appear blue.

in later chapters. Alzheimer's disease becomes worse gradually over years and, as it does, the vast communication system composed of billions of nerve cells is severely disrupted. People who suffer from the disease become permanently unable to remember or perform even simple activities of daily living, like feeding themselves and bathing. Eventually, these people need total care. Unfortunately, there is no way to diagnose Alzheimer's disease with certainty until after death, when brain tissue can be safely examined (Figure 2.1).

Alzheimer's disease affects about 4.5 million people in the United States, according to the National Institute on Aging. Most of the people affected are over 60 years old, and the risk rises with age. However, Alzheimer's disease and other irreversible dementias are not a normal part of aging.

As the average age of our population rises, Alzheimer's dementia is becoming more common. Younger people, between the ages of 30 to 60 years old, can also be affected by a rare form of the disease that might be inherited, called familial Alzheimer's disease. In the more common form of Alzheimer's that affects the elderly, there is no obvious family pattern.

Alzheimer's disease was named after Dr. Alois Alzheimer, who noticed unusual clumps and tangled fibers (Figure 2.2) in the brain of a woman who died of a brain disorder in 1906. Later, scientists discovered that the clumps, now called **plaques**, are made of **beta amyloid**, a toxic molecule that is formed from normal protein. Another protein called **tau** normally helps support neurons, but in cases of Alzheimer's disease, it becomes tangled, disabling the neurons. Gradually, the brain atrophies (shrinks) as more and more neurons die. As these brain changes occur, the symptoms of memory loss, language deterioration, restlessness, confusion, poor judgment, and mood swings progressively worsen (see "An Alzheimer's Case Study" box).

Dementia of the Lewy body type is sometimes confused with Alzheimer's disease, and like Alzheimer's disease it is characterized by an abnormality in brain protein. Lewy bodies are abnormal inclusions in cells found in the neocortex, limbic cortex, subcortical nuclei, and brain stem. A protein known as **alpha-synuclein** appears to be the main component of the Lewy body. In addition to the formation of Lewy bodies, abnormalities in neurotransmitters and loss of nerve cells occur. Like Alzheimer's disease, dementia of the Lewy body type causes a progressive decline in thinking that is severe enough to interfere with normal social function or employment. People suffering from Lewy body type dementia may also have sharp variations in alertness or attention. They may complain of detailed, recurrent visual hallucinations and have repeated, unexplained falls.

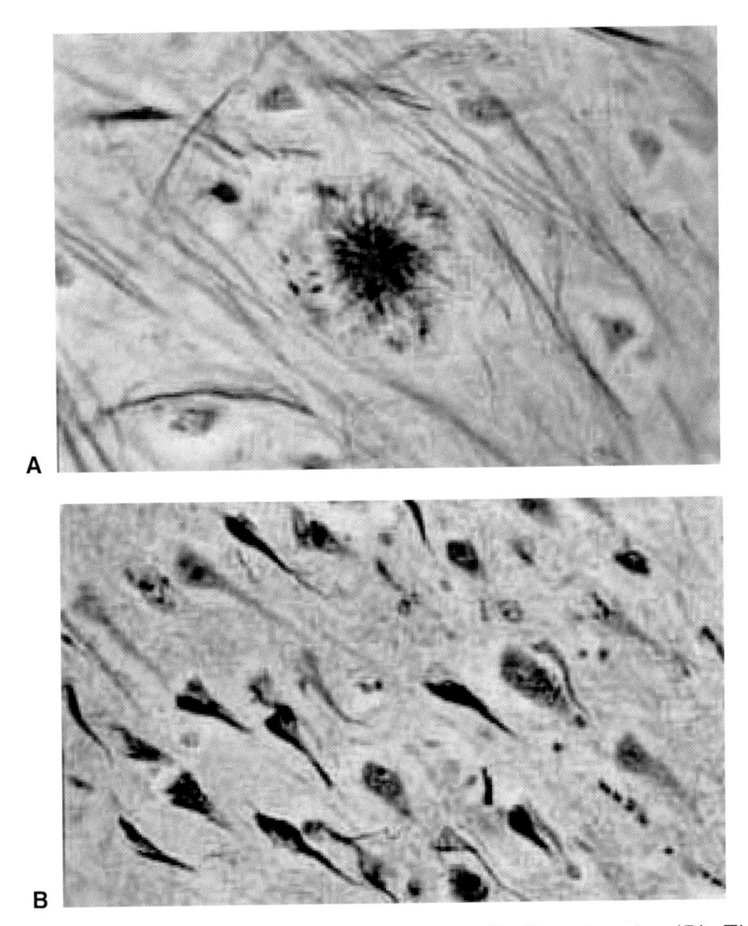

Figure 2.2 Amyloid plaques (A) and neurofibrillary tangles (B). These proteins are found in the brains of patients with Alzheimer's disease.

Neuroscientists are trying to learn more about the risk factors and causes of Alzheimer's disease and Lewy body type dementia. There is no cure for either of these types of dementia, but progress is being made in research (see "A Recent Discovery" box.)

Scientists study the genetics, education, environment, diet, and habits of people with Alzheimer's disease to learn what role these factors might play in the development of this dementia and what strategies might slow its course. Scientists suspect that the same lifestyle risk factors that are bad for the heart, such as lack of exercise, obesity, high blood pressure, and high cholesterol, may also be bad for the brain. Some risk factors specific to Alzheimer's disease include older age (the number of people with the disease doubles every 5 years after age 65) and television viewing. A study at Case Western Reserve University in Cleveland found that each 1-hour increase in daily TV viewing in midlife corresponded to a 1.3 times increased risk of developing Alzheimer's later in life. Television viewing may replace more stimulating social and intellectual activities that may help

An Alzheimer's Case Study

A 60-year-old lawyer began to have short-term memory lapses. She had been practicing law for nearly 30 years. Her colleagues noted that she was "not as sharp" in the courtroom, that she frequently lost her train of thought, and that she misplaced client files and memos. She could not complete her income tax filing. Eventually, she was forced to retire from her law firm because she could no longer adequately represent her clients. Soon after, she was diagnosed with Alzheimer's disease by a neurologist. Her husband hired a home health care worker to assist her with activities of daily living while he was at work. The patient gave up her driver's license at her husband's insistence when she repeatedly missed the correct freeway exit for her home. As her condition progressed, she became agitated and more confused, especially after dark, and yelled at her husband, accusing him of stealing her money. A variety of medications were tried, but none had a significant impact on her memory or functioning. She was no longer able to socialize with friends or family, would wander away if left alone outdoors, and became completely dependent on her home aide. At age 70, the patient contracted influenza, which progressed rapidly to pneumonia, and she died.

to ward off the beginning of Alzheimer's disease. Therefore, general prevention strategies for Alzheimer's may involve including more fruits and vegetables in the diet, reducing saturated fat intake, reducing the number of hours spent watching TV, and maintaining a mentally and physically active lifestyle. A genetic risk factor has also been identified, which involves **apolipoprotein E**, or apoE. ApoE helps carry cholesterol in the blood. One variation in the apoE gene protects against Alzheimer's, and one seems to make an individual more susceptible to the disease. Researchers believe there may be more genes that have an impact on the risk for Alzheimer's. These genes remain to be discovered.

Current research aims to identify those people with early cognitive impairment and then follow their conditions over several years. During that time, these individuals undergo special brain

A Recent Discovery

Researchers recently identified a brain protein that stops the progression of Alzheimer's disease in human brain tissue. *Transthyretin* was first found to protect mice with high levels of toxic beta-amyloid protein in their brains. When human brain cells from the cerebral cortex were grown in culture together with the transthyretin, and then exposed to the toxic amyloid, the brain cell death was minimal. Dr. Jeff Johnson at the University of Wisconsin said that this research could result in a new approach to treatment that is "focused on preventing the loss of brain cells instead of treating the resulting symptoms."[1]

[1] Stein, T. D., N. J. Anders, C. DeCarli, S. L. Chan, M. P. Mattson, and J. A. Johnson, "Neutralization of Transthyretin Reverses the Neuroprotective Effects of Secreted Amyloid Precursor Protein (APP) in APP_{Sw} Mice Resulting in Tau Phosphorylation and Loss of Hippocampal Neurons: Support for the Amyloid Hypothesis." *Journal of Neuroscience* 24 (September 2004): 7707–7717.

imaging, and "biomarkers" in their blood, urine, and cerebrospinal fluid are examined. The goal is to identify the best markers that signal the progression of full-blown Alzheimer's so that the disease can be treated before serious damage to the basal forebrain has occurred.

■ **Learn more about dementia** Search the Internet for *Alzheimer's disease*, *multi-infarct dementia*, or *Lewy body type dementia*.

3 The Cells of the Forebrain

The human nervous system is composed of billions of nerve cells. These nerve cells can be divided into two main types, **neurons** and **neuroglial cells**. In this chapter, we will take a close look at the structure of each of these cells, their function, and their unique contribution to the nervous system.

Neurons are cells that are specialized for communication with other neurons. They receive, conduct, and transmit electrochemical signals. Each neuron has a **cell body** or **soma**, which is the metabolic center of the cell. The soma contains the typical components of a cell, the most important of which is the nucleus. The nucleus contains the genetic material of the cell and a **cytoplasm** made of clear fluid. Neural cytoplasm also contains mitochondria, microtubules, and neurofilaments. **Mitochondria** are involved in the production of energy for the cell. **Microtubules** help to provide transportation routes for molecules within the neuron. **Neurofilaments** provide a supportive matrix.

Surrounding the soma are **dendrites**, which are short, branch-like extensions. The **axon** is one long, single extension from the cell body that ends in many axon branches. At the end of these branches are many terminal areas, called **buttons**, because they resemble buttons (also called bouton terminals). The dendrites receive signals from other

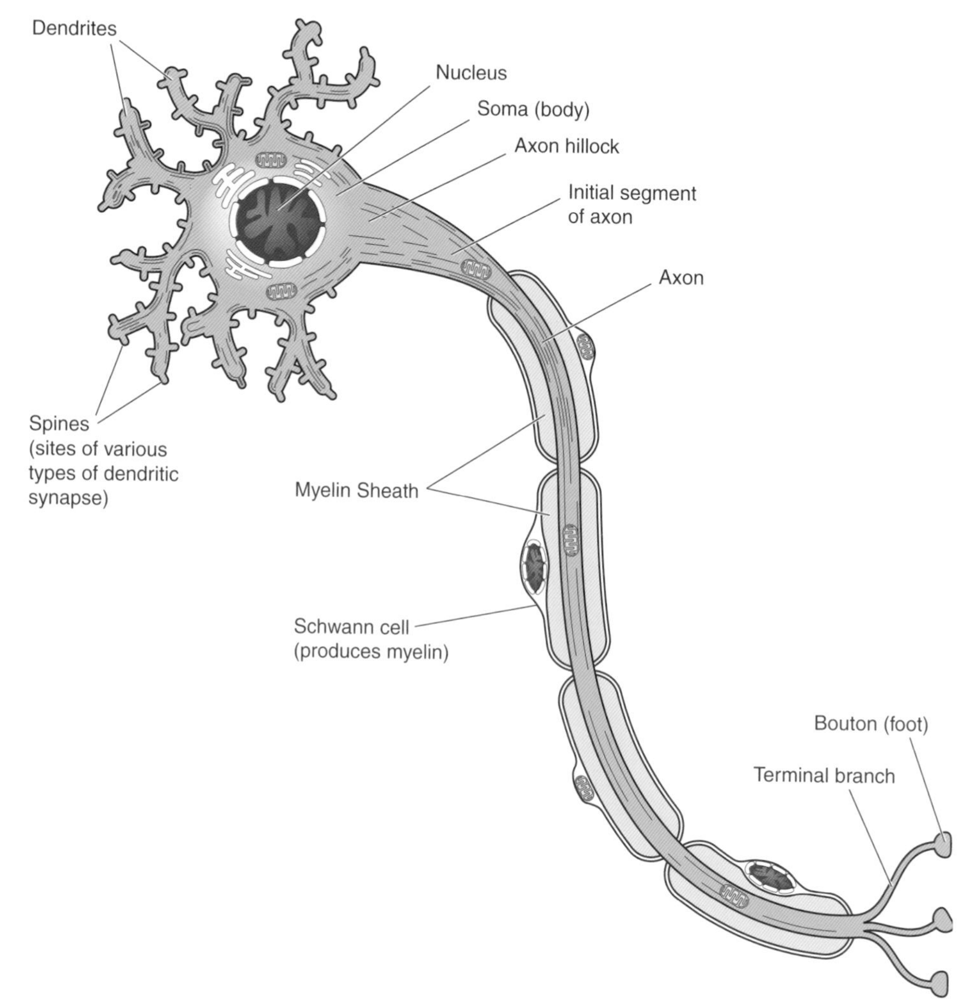

Figure 3.1 Neurons are specialized cells that communicate with other neurons. A neuron consists of the main cell body called the soma, branch-like extensions called dendrites, and a single long extension called an axon. The dendrites receive signals from other neurons, while the axon transmits signals away.

neurons. Sometimes the cell body also receives signals from other neurons. The axon conducts signals away from the soma and transmits them to other cells, usually by way of the buttons (Figure 3.1).

THE RECEPTIVE AREA OF THE NEURON

Input to the neuron comes largely from the buttons of other neurons that connect along the dendrites and the soma. There is a small gap between the buttons of one neuron and the receptive membrane of another neuron that is adjacent to it. This gap, known as a **synapse**, is very important to the functioning of the nervous system. All of the axon's buttons form the pre-synaptic side of the synapse, whereas the dendrites of a receiving neuron form the post-synaptic side. The space between the pre-synaptic and post-synaptic sides is called the **synaptic cleft**. Electron microscopy makes it possible to see a group of membranous sacs, called **synaptic vesicles**, on the pre-synaptic side next to the cleft. **Neurotransmitters** are packaged in these vesicles.

Neurotransmitter molecules are released at the synapses and they influence other cells. Special small membrane sacs called synaptic vesicles store packets of the neurotransmitter molecules until they are ready for release at the **pre-synaptic membrane** of the button. There are many kinds of neurotransmitters, and they range in complexity from simple amines to complex polypeptides. Some neurotransmitters of the central nervous system include acetylcholine, serotonin, epinephrine, norepinephrine, and **dopamine**. A particular neuron usually releases only one or two kinds of neurotransmitters at all of its synapses. Some neurotransmitters *excite* other cells, whereas others *inhibit* them.

What does the release of the neurotransmitter from the synapse do? It causes a small electrical charge on the **post-synaptic membrane**. If the charge excites the membrane, it is called an **excitatory post-synaptic potential (EPSP)**. If, instead, the charge inhibits the membrane, it is called an **inhibitory post-synaptic potential (IPSP)**. Both EPSPs and IPSPs become weaker as they travel past the post-synaptic membrane and eventually stop part way down the axon. The function of the EPSPs and

IPSPs is to affect other electrochemical signals that don't die out.

These EPSPs and/or IPSPs travel to the **axon hillock**, a cone-shaped junction between the cell body and the axon. This area is the neuron's trigger zone. The axon hillock adds all the EPSPs and IPSPs that reach it, and if the level of excitation exceeds the level of inhibition, then the threshold of excitation is reached. When the threshold of excitation is reached, then an **action potential** is produced. Action potentials then travel down the axon to the terminal buttons.

Action potentials are "all or none." In this way they differ from EPSPs and IPSPs, which vary in strength. Also, EPSPs and IPSPs are conducted decrementally along the axon, which means that they become weaker as they travel. Action potentials reach the terminal button as large and as strong as when they left the axon hillock.

What does an action potential do? It triggers the release of some of the neurotransmitter molecules from storage in synaptic vesicles near the pre-synaptic membrane. The synaptic vesicles bind to the pre-synaptic membrane and split open, releasing the contents into the synapses. This process of neurotransmitter release is called **exocytosis.** When the vesicle attaches to the pre-synaptic membrane, it becomes part of it.

The released neurotransmitters then cross the synapse and bind to receptors in the post-synaptic membrane. Each type of neurotransmitter usually has a specific receptor. When the neurotransmitter binds to the receptor, it usually induces an EPSP or IPSP in the post-synaptic membrane. A particular synapse is either excitatory or inhibitory, not both.

Neurons are like links in a chain that are lined up but separated from each other at the synapses. Neurotransmitters travel across the synapse between the terminal buttons of one neuron and the dendrites of the next neuron. The dendrites transform

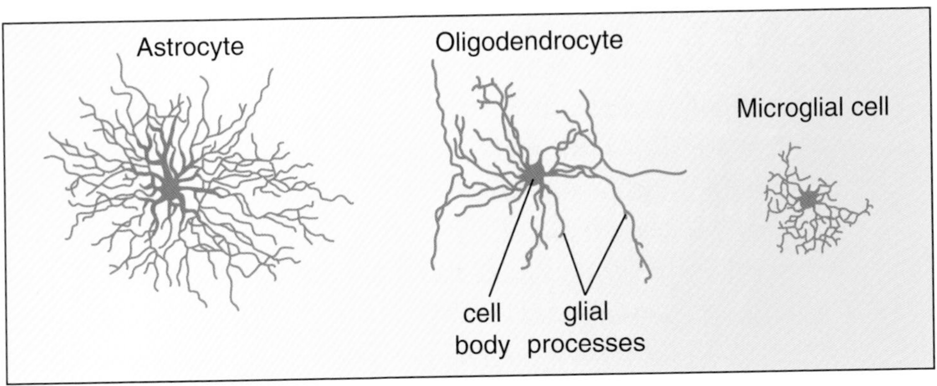

Figure 3.2 The major types of glia in the central nervous system are called astrocytes, oligodendrocytes, and microglia.

the charge they receive into EPSPs and IPSPs, which travel down the neuron to the axon hillock. At the axon hillock an action potential is produced. The action potential travels to the terminal buttons and stimulates the release of neurotransmitters, which cross the synapse to the dendrites of the next neuron, and so the process begins over again.

NEUROGLIAL CELLS

There are about 100 billion neurons in the brain, and many more neuroglial cells. Neuroglial cells are responsible for about 40% of the total volume of the brain and spinal cord. Neuroglial cells are also called glial cells, or simply glia (Figure 3.2). There are different kinds of glial cells, each of which performs a different supportive or protective function in the nervous system. Some glial cells hold neurons in place with a physical framework. Some remove foreign matter and other debris, and some may regulate the blood–brain barrier (see "The Essential Glial Cell" box).

In the central nervous system (CNS), the neuroglia include microglia, macroglia, and ependymal glia. In the peripheral nervous system (PNS), the neuroglia include **Schwann cells** and satellite cells. The neuroglia differ from neurons because they can potentially divide after birth and may form tumors of the brain that are slow-growing. They do not form synapses, and they cannot transmit impulses. However, they can participate in certain electrical phenomena.

Macroglia include astrocytes and **oligodendrocytes**. Astrocytes are the most common glia of the CNS (Figure 3.3). They are subdivided into protoplasmic astrocytes and fibrous astrocytes. Protoplasmic astrocytes have a branching form, whereas fibrous astrocytes have mostly unbranched processes. Both types of astrocytes are larger than oligodendrocytes.

Microglia, the smallest neuroglia, proliferate in response to injury in the CNS. They scavenge and ingest foreign matter, and when they do so they become distended. In their distended state, they are referred to as "glitter cells." Ependymal cells line the

The Essential Glial Cell

Glial cells do more than support neurons. According to researchers at Stanford University School of Medicine in California, glial cells are essential in the development of synapses. Glial cells make two proteins, called *thrombospondins*, that signal synapse formation. Mice that have been genetically engineered to lack thrombospondins produce 40% fewer synapses on average compared with normal mice. Researcher Dr. Ben Barres said, "We knew glia had a close relationship with neurons. We never thought that synapses would entirely fail to form without the glia."[1] The researchers believe that these findings may lead to improved care for patients with brain damage, epilepsy, and addiction.

[1] Christopherson, K., E. M. Ullian, C.C.A. Stokes, et al. "Thrombospondins Are Astrocyte-Secreted Proteins that Promote CNS Synaptogenesis," *Cell* (2005): 421–433.

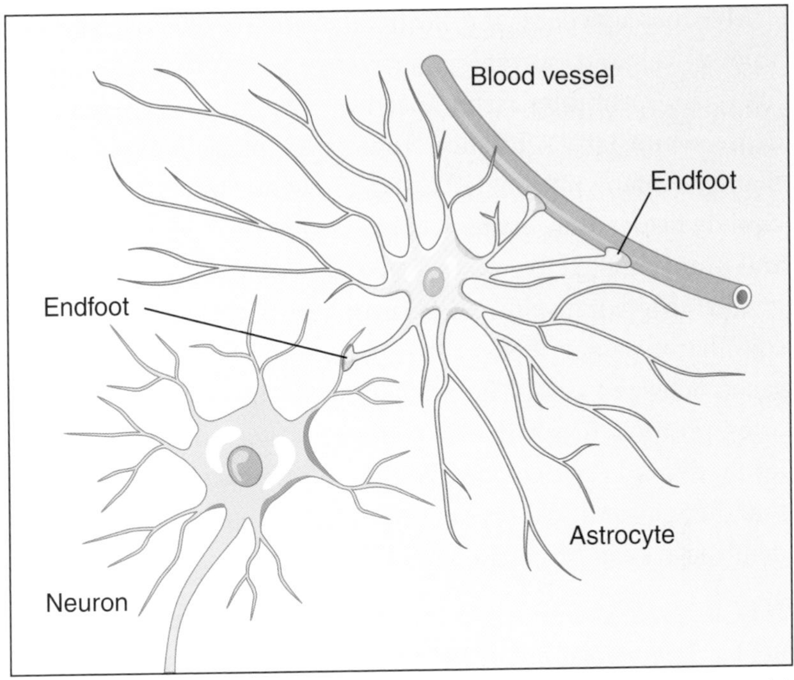

Figure 3.3 One role of astrocytes in the brain is to supply neurons with the nutrients they need to stay alive and active. To do this, astrocytes pick up the nutrients from the blood vessels with special contacts called endfeet and shuttle them to the neurons.

cavities of the brain (in the ventricles) and spinal cord (in the central canal) in a single cell layer.

Some glial cells produce **myelin**, which wraps axons, or myelinates them. Gaps called **nodes of Ranvier** are present along the axon. At the nodes of Ranvier there is no myelin. These nodes are the only places on the axon where ionic current can flow. In myelinated axons, action potentials quickly jump from node to node and do not travel at a constant speed along the axon. Saltatory (jumping) conduction in myelinated axons increases the speed of axonal conduction. In large myelinated axons, action potentials are conducted along the axon at about 100 meters per second (for comparison, sound travels at 340 meters per second).

Myelin is a physical and electrical insulator, like insulation on a power cable. It prevents interference between adjacent nerve fibers and promotes faster speed of transmission. Myelin is a dense, white, fatty substance. Those areas of the nervous system that have many myelinated axons are white and are referred to as **white matter. Gray matter** is composed of cell bodies, dendrites, and unmyelinated axons.

Glial cells that myelinate CNS axons are oligodendrocytes. Glial cells that myelinate PNS axons (outside the skull and spine) are called Schwann cells. The Schwann cells, but not oligodendrocytes, promote regrowth or regeneration of axons in the PNS after they have been damaged. Another difference between them is that the Schwann cells myelinate one fiber at a time, but the oligodendrocytes can myelinate several fibers simultaneously.

■ **Learn more about cells of the nervous system** Search the Internet for *neurons* and *neuroglial cells, neurotransmitter*, or *glial cell.*

4 Development and Organization of the Brain

Human brain formation begins very early in prenatal life. During the first five months of pregnancy, the fetus forms about 100 billion neurons. In the first and second trimesters (the first and second three months of pregnancy), synapses are formed at a rate of about two million every second.

Our brain is a continuous work in progress. For many years, scientists believed that new neurons were not produced after the human brain reached adult size, but recent evidence suggests that new neurons are produced throughout life, in a process called **neurogenesis**, though probably in a smaller number and at a slower pace from the brain's initial formation (see "Neurogenesis" box).

Brain size increases gradually as the individual neurons grow and sprout hundreds of dendrites (Figure 4.1). In fact, most of the growth of the brain (by volume and weight) is due to the growth of dendrites. A newborn baby has about 25% the brain size of an adult, but by 3 years of age, the size of that child's brain will increase to about 80% of an adult brain and to 90% by age 5.

Brain development begins with processes called **gastrulation** and **neurulation**. Gastrulation occurs first. During this process, the developing embryo undergoes an elaborate series of infoldings that results in the formation of the **neural tube**, which grows along the back of the human embryo (Figure 4.2). This is a very sensitive and important time in development. At about 16 days after conception, a patch of special cells called the **neural plate** appears on the **dorsal** surface. This plate lengthens and folds upward to form a groove at about 18 days. The two edges of the groove then fuse to form a fluid-filled neural tube, which eventually becomes the **central nervous system** (**CNS**). As the tube

Neurogenesis

For over a century, it was believed that after infancy, neurons do not divide or grow and that we are born with every brain cell that we will ever have. The brain was considered the only organ of the body that could not produce new tissue to heal an injury. That is how scientists explained the fact that patients rarely recover completely from strokes and that strokes have such damaging and permanent impacts.

Then, neuroscientist Fernando Nottebohm of Rockefeller University began to investigate why canaries were able to learn new songs. Nottebohm was interested in bird song because he considered it the closest thing in the animal kingdom to human speech. Many types of birds sing just one song for the duration of their lives, but male canaries, which live for an average of 10 years, are able to learn a new song each year. Male canaries sing a particular song throughout the spring breeding season. The following spring they have learned a new song. In 1981, Nottebohm developed a theory that the neurons in charge of the old bird song were dying, and new neurons were being born. Using injections of radioactive tracer molecules, Nottebohm determined that canary brains produced thousands of new neurons each day. But, because the human brain's inability to produce forms, cells from the neural plate break off and move lateral to the tube, forming the **neural crest**, which develops into the **peripheral nervous system** (**PNS**). The tube is fully closed by 27 days and has begun its transformation into the brain and spinal cord (see "**Neural Tube Defects**" box).

In the fourth week of prenatal growth, three swellings become visible in the **anterior** end of the neural tube (the part which will

new neurons was such a widely held belief, scientists were skeptical about his results.

Later, a young psychologist currently at Princeton University, Elizabeth Gould, wanted to explain why rats that had had their adrenal glands removed, resulting in the death of many cells in the hippocampus, ended up with as many cells in the hippocampus as healthy rats. If many cells were dying, how could there be the same number as were expected in healthy rats? She determined that neurogenesis was the only possible answer. Gould then demonstrated that neurons are not only born in the hippocampus of adult rats, but also in the hippocampus of adult tree shrews. Later, she demonstrated neurogenesis in marmosets and macaques. Each animal brain she studied was a step closer to being like the human brain.

In 1998, by studying the brains of cancer patients who had received injections of a tracer chemical, Fred H. Gage of the Salk Institute in La Jolla, California, demonstrated that new cells can grow in the hippocampus of human adults.

Neurogenesis is still a new field, and the claims of these scientists remain controversial, but more and more work is being done to add to their findings.[1]

[1] Specter, M. "Rethinking the Brain: How the Songs of Canaries Upset a Fundamental Principle of Science," *The New Yorker* (July 15, 2001): 42–53.

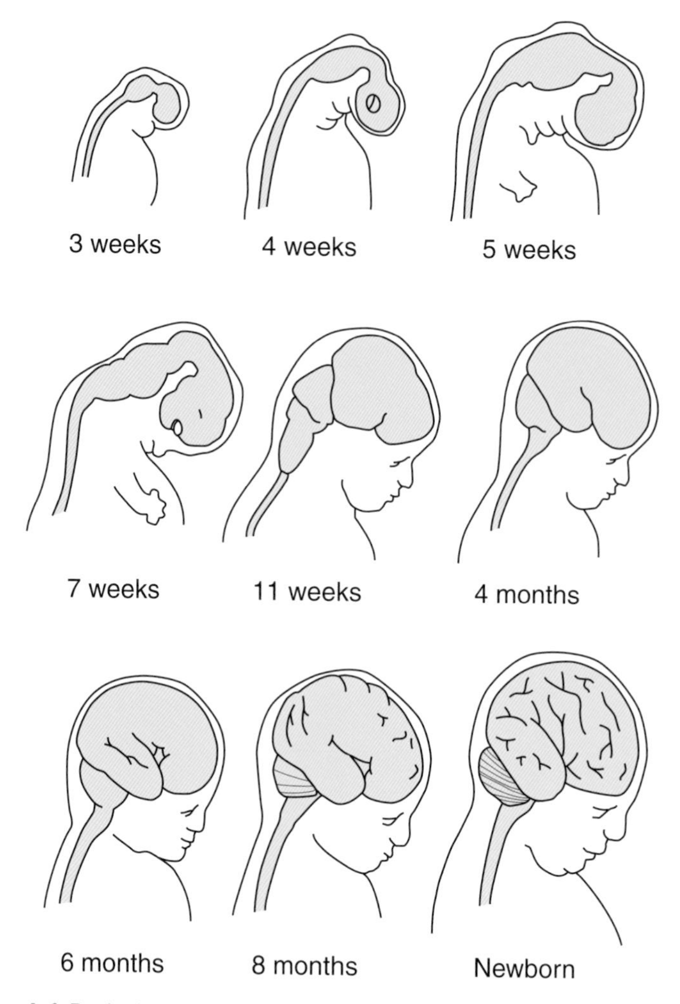

Figure 4.1 Brain formation begins early in the growth of the human embryo. By the fourth week, structures of the hindbrain, midbrain, and forebrain have started to develop.

become the brain). These three swellings are the forebrain, the midbrain, and the hindbrain. The forebrain is the **prosencephalon**, the midbrain is the **mesencephalon**, and the most posterior of the three swellings is the **rhombencephalon**. The rest of the neural tube develops into the spinal cord.

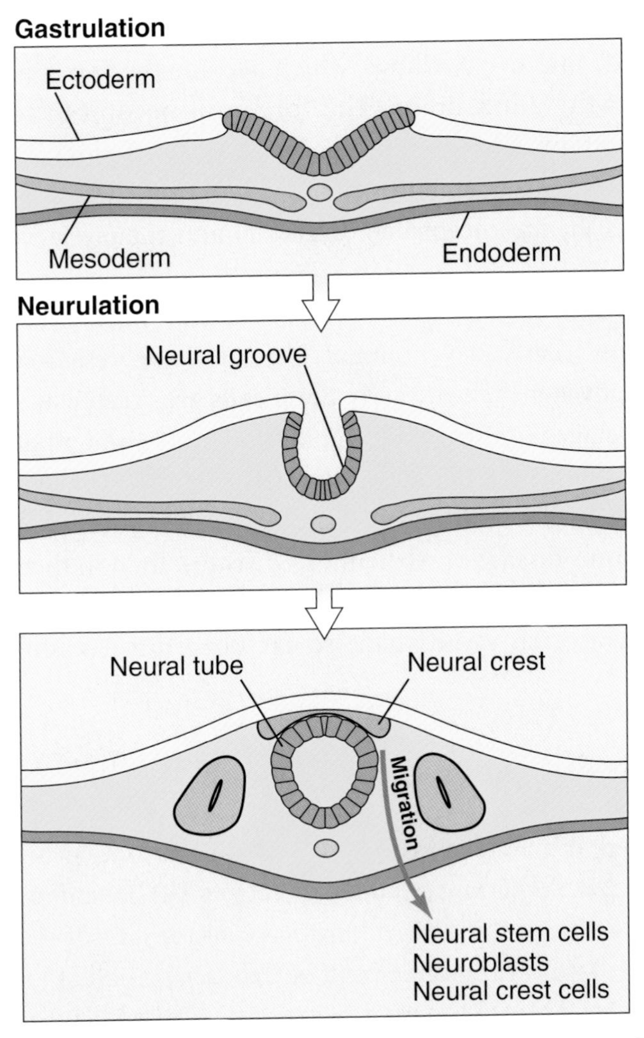

Figure 4.2 Nervous tissue is generated during gastrulation and neurulation, during which the embryonic neural plate is formed, transformed into the neural groove, and then transformed into the neural tube. Migration of neuronal precursors from the neural tube is an important event in this development.

In the fifth week of prenatal life, the three swellings further subdivide into five swellings, which become the five major divisions of the adult brain. The forebrain becomes the telencephalon (anterior) and the diencephalon (posterior). The midbrain, or mesencephalon, stays the same. The hindbrain becomes the **metencephalon** (anterior) and the **myelencephalon** (posterior).

An area of interest to many neuroscientists that is related to the brain's early development is fetal tissue research, more specifically stem cell research. **Stem cells** are created at the very earliest stages of embryonic life and have the potential to develop into any type of cell in the body. If they can be induced to replace cells damaged by certain brain disorders (such as Parkinson's disease or Alzheimer's disease), then in theory they could be grafted into a human brain to repair the damage. Although research results have so far been unsuccessful, some

Neural Tube Defects

If the neural tube fails to close during the fourth week after conception, two important neural tube defects (NTDs) can occur. At the brain end of the embryo, *anencephaly*, or lack of a cerebral cortex, is fatal. At the lower end of the tube, a defect known as spina bifida occurs. In *spina bifida*, part of the spinal cord develops outside the spine. The symptoms of spina bifida range in severity but can include loss of bowel or bladder function, sensory loss, and paralysis. About 60% of NTDs can be prevented by the vitamin *folic acid*. A public health effort in the United States recommends that women trying to become pregnant take 400 micrograms of folic acid supplement daily, beginning about 1 month before conception and continuing through the first trimester (3 months) of pregnancy.

scientists believe that, eventually, human stem cells could be grown in a lab and transplanted into the brain of a person with brain damage in order to repair that damage. Stem cell science uses embryos created *in vitro* ("in glass"), in other words embryos fertilized in medical clinics. Stem cell research is a controversial topic because some people think that life begins with fertilization and so are opposed to research being performed on embryos.

Many neurons are created after the neural tube is formed. These neurons then migrate outward to appropriate locations in the neural tube and aggregate to form specific neural structures. Later, axons and dendrites grow from the developing neurons and form synapses. It is currently believed that all the neurons in the cerebral cortex are produced before birth, but they are poorly connected. Let's consider the development of the cortex in more detail. The early telencephalic wall contains a rapidly growing group of neuroblasts that will become neurons later on. These neuroblasts move up the brain wall in successive waves, like amoebas, through the groups of cells that are already in place and undergoing further differentiation. Therefore, the deepest layers of cells reach their intended position first. The six layers of the cerebral cortex eventually form in this manner. Neurons and glia proliferate very rapidly in these early stages.

The cerebral cortex forms most of its synaptic connections after birth, during the **exuberant period**. By the time a toddler is 2 years old, his or her cortex will contain more than 100 trillion synapses. The number of synapses in the brain peaks early but will drop to about one-third of this level between early childhood and adolescence and will continue to decline until adulthood. The process by which synapses are eliminated is called **pruning**. Pruning streamlines neural circuits by eliminating unused synapses, thus increasing efficiency and processing speed in the brain. In addition to synapse formation and pruning, myelination also occurs. While the brains of newborns contain little myelin, myelination begins within the first two years of life and continues through childhood and adolescence. In the cerebral cortex, myelination begins in the primary and sensory areas and progresses to association regions, which control the complex integration of memories, emotions, perceptions, and thinking.

Myelination requires fat in the diet, so poor nutrition in early childhood can compromise brain development. The quality of a pregnant woman's diet is very important for optimal fetal growth and development in general. Malnourished children have abnormally small brains because of reduced myelination and fewer glia. Also, these children are likely to have reduced dendritic growth. As a result of these deficits, malnourished fetuses and infants can have permanent deficits in their behavior and cognition (ability to think). These deficits show themselves as learning disorders, low academic achievement, lower IQ, and slow language and fine motor development.

One common cause of serious behavioral and intellectual problems in children is fetal alcohol syndrome. In this condition, the developing fetus is exposed to alcohol in the bloodstream of the mother, and the alcohol damages the fetal brain. Also, an alcoholic mother may substitute alcohol for more nutritious, healthy foods, which further damages the unborn baby. Sometimes a woman may drink heavily during the most vulnerable brain-development period for a fetus—the first trimester or 12 weeks—without realizing she is pregnant. Some children with fetal alcohol syndrome are born mentally retarded, and some may be hyperactive and difficult to teach.

Of course, alcohol is not the only toxin that can damage the fetal brain. Researchers are also studying the effects of street drugs such as heroin, amphetamines, marijuana, and cocaine on fetal brain development and cognitive functioning in later life. Cigarettes, radiation, pesticides, mercury, and many industrial chemicals can also potentially damage the growing fetal brain.

Brain growth continues after birth and so does the need for adequate nutrition. The best nutrition for newborns is human breast milk. Breast milk has the ideal mixture of nutrients for brain growth, including necessary fats. Some recent studies have shown that breast-fed babies have slightly higher IQs compared to formula-fed babies. However, the relationship between breast-feeding and higher intelligence is not necessarily causal. It is important for researchers to be able to sort out variables that influence brain development, such as maternal intelligence and education and an enriched environment.

What parts of the baby's brain mature first? Those structures responsible for survival outside the womb, such as the brain stem, will be needed first if a baby is born premature (before it has spent 40 weeks in the uterus of its mother). The brain stem controls vital functions such as heart rate, breathing, and blood pressure. By six months of uterine life, a baby's brain stem is mature enough to take on these independent functions. The last part of the brain to mature is the cerebral cortex, the most sophisticated area of the human brain, which is responsible for feeling, thinking, voluntary actions, and remembering. The development of the cortex occurs more slowly in contrast to the lower portions of the brain, including the brain stem and the spinal cord. These control fundamental actions such as reflexes, rooting (the baby turning toward the nipple to breast-feed), sucking, feeding, crying, and sleeping. Babies are very good at letting their needs be known to their caretakers via loud crying since speech and language have not yet developed. Visual circuits within the brain stem are probably responsible for babies' ability to follow or "track" their parents' faces.

Babies are "born to learn," and they seek information through all their senses. Babies elicit protection and nurturing from their parents in the form of holding, rocking, stroking, singing, and talking. All of these actions provide stimulations needed by the growing brain. In this sense, infants are active participants in their own development.

The "executive" part of the brain, the prefrontal cortex, matures at about age 25. Executives are managers, and managing is the role of the **prefrontal cortex:** it sets priorities, controls behavior, makes social judgments, and plans ahead.

The maturity of gray matter in adult life does not mean that the brain remains static after that. **Plasticity**, or changes in the connections between neurons as a result of experience, occurs in both babies and the adult human cortex. This process is especially important in memory formation (a complex process that we will explore later in this book) and learning. Neuroimaging studies reveal that musicians have more sensory cortex related to their instrument-playing fingers compared to non-musician subjects. The enlarged cortical areas may indicate that highly used senses result in greater numbers of neurons in these brain areas and possibly in enhanced sensitivity in the fingers during music playing as well. Better understanding of plasticity could improve recovery efforts for those people who suffer from sensory deficits as a result of strokes or injury to the brain.

■ **Learn more about the evolution of the brain** Search the Internet for *neurogenesis, evolution and brain,* or *embryonic brain.*

5 Structures of the Forebrain

As you recall from the previous chapter, the forebrain develops in the fifth week of prenatal life into the telencephalon and the diencephalon. The two cerebral hemispheres together compose the telencephalon, which is the biggest division of the human brain. This part of the brain is also known as the **cerebrum**. The cerebrum sits above the vertebrate brain stem. The brain stem is made up of the other divisions of the brain. From anterior to posterior, these are the diencephalon, the mesencephalon, and the telencephalon. The diencephalon is composed of the hypothalamus and the thalamus, which are in turn composed of many pairs of nuclei. Nuclei (singular: nucleus) are structures composed mostly of cell bodies, found in the CNS. Their counterparts in the PNS, also composed mostly of cell bodies, are called ganglia (singular: ganglion). These collections of cell bodies perform local analysis of neural signals. They are not to be confused with the other "nucleus," the spherical structure within the cell body of the neuron.

The forebrain is thus primarily composed of the cerebral hemispheres, the hypothalamus, and the thalamus. The hypothalamus and thalamus are discussed in greater detail in Chapter 5, and the cerebral cortex, because it is the most substantial part of the forebrain, is discussed at length in its own chapter (Chapter 6). Two other systems of the cerebrum, the limbic system and the basal ganglia, are also discussed in later chapters (Chapters 7 and 8). This chapter is devoted to other important structures of the forebrain, including blood vessels, **meninges**, cranial nerves, and **commissures.**

BLOOD SUPPLY

The brain is a very active organ that requires a continuous flow of blood to maintain its normal activity. The brain is only able to store very small amounts of glucose and oxygen, and therefore, it depends almost entirely on aerobic metabolism of glucose in the bloodstream. The brain uses about 20% of the total oxygen consumed by all the body's tissues. The brain receives about 800 milliliters of blood flow every minute. If blood supply is cut off for 10 seconds or less, a person will become unconscious. If blood flow is interrupted for 5 to 10 minutes, irreversible damage to the neurons can occur. The brain has two main supplies of arterial blood, which come from the vertebral arteries and the internal carotid arteries. The cerebral hemispheres receive blood from both the internal carotid arteries, and the anterior, middle, and posterior cerebral arteries as well. The deep, or basal, forebrain receives its blood supply from deep branches of the cortical arteries. This area involves the basal ganglia and the diencephalon (thalamus and hypothalamus). The blood supply to these areas of the deep forebrain is critical, as a stroke (a blood clot) or hemorrhage here can cause severe disability and/or death. The arteries involved include the anterior cerebral artery, the medial striate arteries, the anterior communicating arteries, the middle cerebral artery, and the lenticulostriate arteries. The lenticulostriate arteries are often the site of a stroke or hemorrhage, because they are thin and small. If these arteries are damaged, the motor pathways in the internal capsule (made up largely of fibers connecting the thalamus and the cere-

Figure 5.1. This photo shows a patient with hemiparesis, a weakening or paralysis of one side of the body, which sometimes follows a stroke. Physical therapy can help patients regain the use of their paralyzed limbs.

bral cortex) are deprived of blood, and as a result, paralysis will occur on the opposite side of the body from the affected motor pathways. This is known as hemiplegia, or partial paralysis.

The posterior cerebral artery, the posterior communicating artery, and the thalamic branches supply the posterior thalamus and the posterior limb of the internal capsule. Damage to these vessels may cause hemiplegia (Figure 5.1) and contralateral anesthesia (loss of sensation on the opposite side of the body).

The cerebral hemispheres are drained by cerebral veins, which are of two types. The superficial veins drain the cortical surface, and the deep veins drain the basal ganglia, the diencephalon, and other interior parts of the hemispheres. The veins, in turn, drain into dural sinuses, or channels within the dura mater (one of the meninges). The dural sinuses then drain into the internal jugular vein and emissary veins, which exit the skull.

MENINGES

The central nervous system (CNS, brain, and spinal cord) is covered by three connective tissues or membranes which serve a protective function. The outer membrane is the toughest, called the **dura mater.** It contains strong connective tissue and lies next to the skull. The middle membrane is the **arachnoid membrane,** which resembles a spider web. Between the arachnoid membrane and the delicate **pia mater,** which adheres to and covers the brain and spinal cord, is the **subarachnoid space.** The subarachnoid space is filled with cerebrospinal fluid (CSF), which acts like a support and cushion for the brain, protecting it from the hard surface of the skull. The hollow interior of the brain and the spinal cord are also filled with CSF. When CSF is lost (for example, during brain surgery or when it is drawn for examination in a procedure called a lumbar puncture) people complain of severe headaches when they move their heads, because some of the cushioning effect is temporarily gone (Figure 5.2).

VENTRICLES

As the CNS develops in the embryo, the forebrain region undergoes changes in its CSF-filled core. The core expands into several CSF-filled spaces, which are interconnected. These spaces are called **ventricles.**

There are four cerebral ventricles filled with CSF: two lateral ventricles in each cerebral hemisphere, one ventricle in the midline of the diencephalon, and one ventricle in the metencephalon, which connects the central canal of the spinal cord to

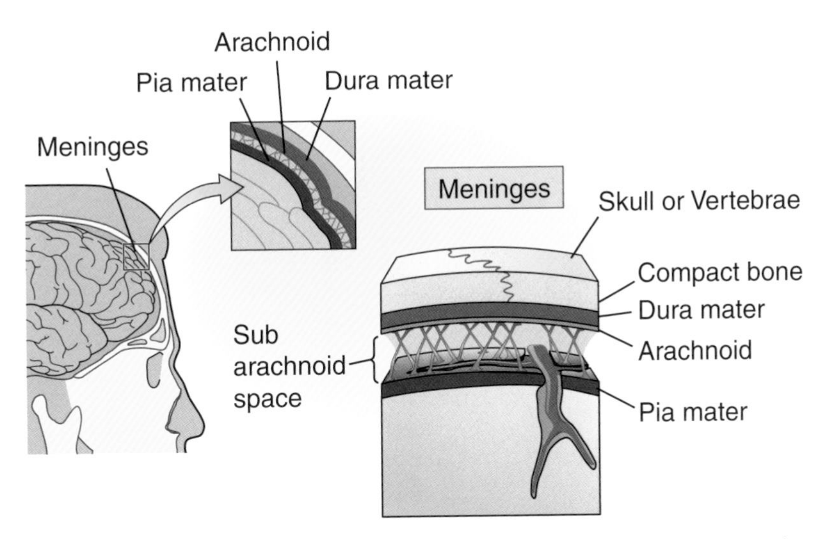

Figure 5.2 The central nervous system is surrounded by three protective layers: the dura mater, the arachnoid membrane, and the pia mater. These layers, called the meninges, support and cushion the brain.

the cerebral aqueduct in the mesencephalon (Figure 5.3; See also "Cerebrospinal Fluid" box).

COMMISSURES

The longitudinal fissure separates the cerebral hemispheres along the **midline**, but the two hemispheres are also connected by tracts, called commissures, that run through the longitudinal fissure. The largest and best known of these is the **corpus callosum**. It contains about 200 million axons.

The **anterior commissure** lies between the left and right temporal lobes and is located **inferior** to the corpus callosum, below its anterior tip. The **massa intermedia** is in the middle of the third ventricle, connecting the left and right lobes of the diencephalon.

The anterior commissure is a major communication pathway between the left and right temporal lobes. The corpus callosum is a wide band of myelinated fibers that connects cortical regions

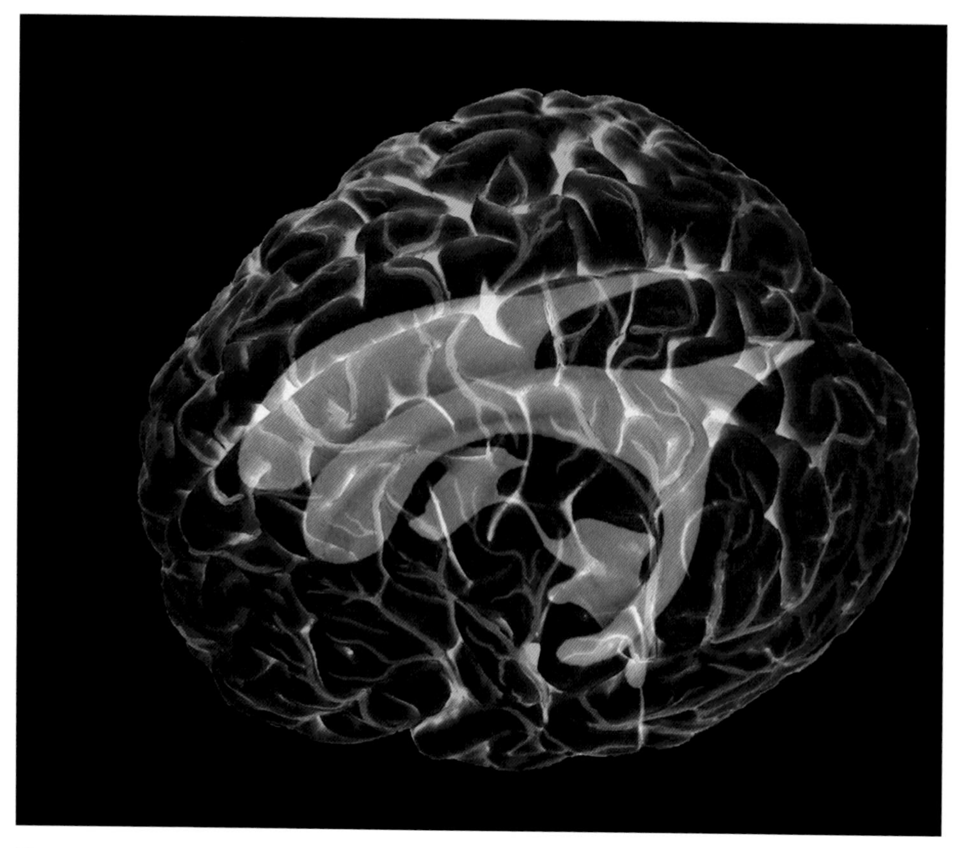

Figure 5.3. The four ventricles of the brain. These ventricles are spaces filled with cerebrospinal fluid.

in one lobe with the same region in the opposite lobe. Cutting the corpus callosum stops communication between the hemispheres, and may help improve severe cases of epilepsy by preventing abnormal signals from traveling (Figure 5.4).

CRANIAL NERVES

Nerves are structures in the PNS and are composed mostly of axons. Their function is to transmit action potentials between parts of the PNS. Most nerves exit from the spinal cord. However, there are 12 pairs of **cranial nerves** that exit directly from the brain. They are numbered in Roman numerals from anterior to posterior. These nerves are extremely important for functions such as smell, vision, eye movement, taste, salivation, swallowing, and hearing. These 12 nerves are considered the peripheral nerves of the brain. Besides the 12 cranial nerves, all other

Cerebrospinal Fluid

What is CSF made of? This liquid is very similar to the part of the blood called plasma. It is a clear fluid with no cells. CSF has less protein and fewer potassium and calcium ions than plasma, but more sodium, chloride, and magnesium ions. It has been described as an ultra filtrate of the blood. The volume of CSF in the average adult brain is about 150 milliliters. CSF is produced in choroid plexuses, a spongy mass of tissue found in the lateral, third, and fourth ventricles. About 700 milliliters of CSF is made every 24 hours, and an equal amount is absorbed back into the venous system. The brain and the spinal cord float in the CSF. The pressure of CSF is lower than blood pressure, and varies with heart rate and respiration.

If CSF has to be tested for signs of infection, or anesthesia has to be introduced (e.g., women who receive epidural anesthesia to block the pain of childbirth), a lumbar puncture is performed at the fourth lumbar level. This puncture is called a spinal tap. A needle is inserted below the end of the spinal cord, and CSF is collected, or anesthesia can be injected by a syringe.

During the development of the human fetus, the normal flow of CSF may become blocked. If too much CSF is produced, or too little is absorbed, hydrocephalus can occur. In this condition, the ventricles swell, and in severe cases the skull is almost completely full of CSF, obliterating the space needed for neurons. Without much brain tissue, babies with hydrocephalus usually die either before or soon after birth.

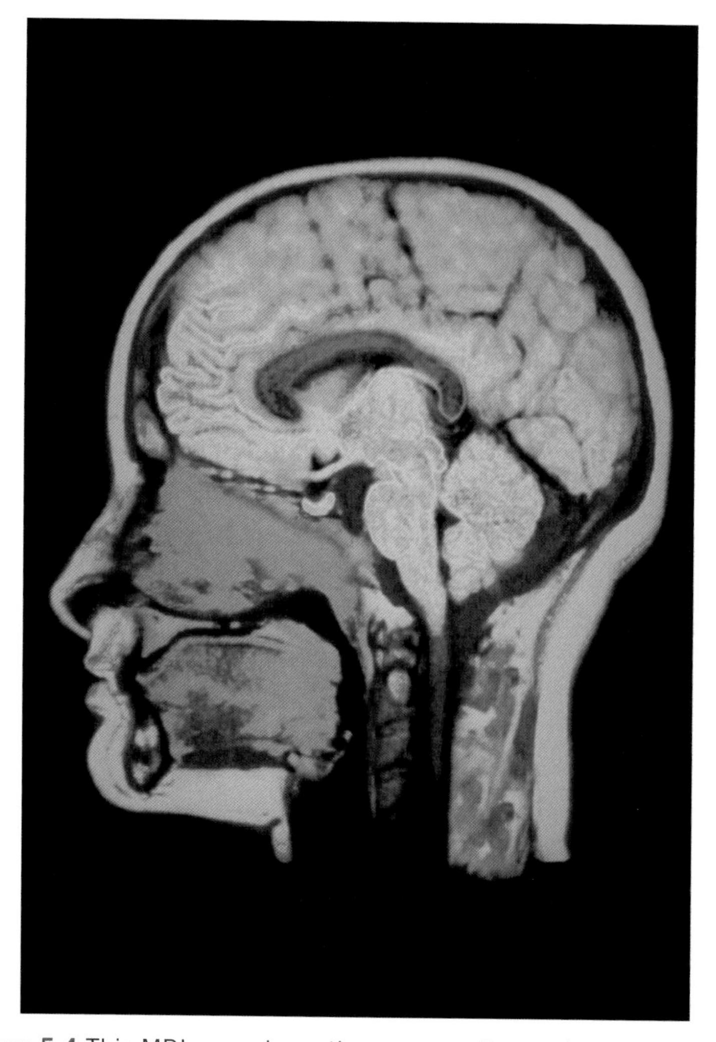

Figure 5.4 This MRI scan shows the corpus callosum in blue. The corpus callosum is the largest of the commissures and enables communication between the brain hemispheres.

sensory and motor nerves enter and leave the CNS through the spinal cord.

Each of the cranial nerves is associated with a specific embryonic brain region.

We will discuss only the cranial nerves associated with the forebrain. These are the olfactory nerve (I), which arises from the telencephalon, and the optic nerve (II), which arises from the diencephalon. The olfactory nerve is composed of short, unmyelinated axons located in the roof of the nasal cavity; they are receptive to smell (olfaction). They carry sensory signals from the receptors of the nose to the brain. The olfactory system is the only sensory system that does not synapse in the thalamus before it reaches the cerebral cortex. Olfaction is a primitive sense and is associated with instincts such as eating, sexual behavior, and procreation. The optic nerve carries sensory signals from visual receptors of the eyes to the brain. The neural impulses transmitted from the eye follow visual pathways to the visual cortex on the medial surface of the occipital lobe. The optic nerve leaves the posterior surface of the eyeball and enters the cranial cavity via the optic canal.

■ **Learn more about structures of the forebrain** Search the Internet for *meninges*, *cranial nerves*, or *commissures*.

6 Functions of the Forebrain

As we learned in previous chapters, the forebrain is composed of the telencephalon and the diencephalon. The telencephalon becomes the cerebral hemispheres, and the diencephalon (part of the brain stem) becomes the hypothalamus and thalamus.

The thalamus is located at the top of the brain stem. It has two large, egg-shaped lobes, one on each side of the third ventricle. Between the two lobes is the massa intermedia, which has been described as both a cerebral commissure and a thalamic nucleus. Most of the thalamus is gray matter, but it has internal layers that contain myelinated axons, giving it a striped appearance (Figure 6.1).

The thalamus is an integrator and hub that is involved in almost all the activities of the forebrain. Most importantly, it relays incoming sensory pathways to the cerebral cortex, allowing us to process sensory information. All sensory pathways have direct projections to and from thalamic nuclei, except for olfactory pathways (relating to the sense of smell). The ventral posterior nuclei relay tactile information (relating to the sense of touch), the medial geniculate nuclei relay auditory information (relating to the sense of hearing), and the lateral geniculate nuclei relay visual information (relating to the sense of sight) to the cerebral cortex. In addition to

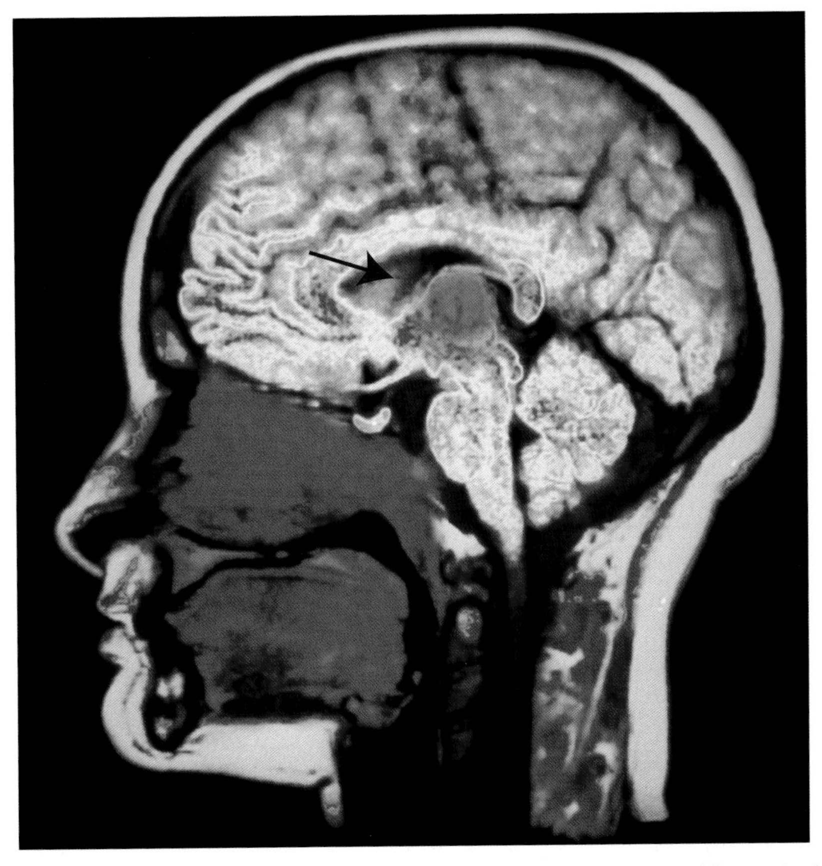

Figure 6.1 The thalamus is located at the top of the brain stem and is marked here with a green dot and an arrow. It is an important hub of activity in the forebrain.

relaying sensory information, the thalamus also plays a role in modulating it.

The hypothalamus is located adjacent to the third ventricle, between the cerebrum and the brain stem and below the thalamus. It has many important functions in **homeostasis**, or the maintenance of a steady state in the body. These functions include regulation of body temperature and water metabolism.

The hypothalamus has many pairs of nuclei. The **ventromedial nuclei** help to regulate the conversion of blood glucose to body fat. The **suprachiasmatic nuclei** contain the biological clock, which regulates the body's 24-hour light/dark cycles, also known as circadian rhythms. Animals (as well as plants) are biologically regulated according to the 24-hour cycle of daylight and darkness. The suprachiasmatic nuclei interpret information from the retina and relay it to the pineal gland. The pineal gland secretes the hormone melatonin (a large amount in the night time, and a small amount during daylight hours). Although the retina is involved, the suprachiasmatic nuclei do not react quickly to changes in the light/dark cycle. This is why travelers experience jet lag. The **preoptic area** of the hypothalamus plays a role in sexual behavior. The mamillary bodies play an important role in emotional regulation.

■ **Learn more about the brain's biological clock** Search the Internet for *circadian rhythms* or *jet lag*.

The hypothalamus also serves the role of regulating the pituitary gland. The pituitary gland has two parts, anterior and posterior, and it releases body hormones. It is a small, spherical gland located in a bony cavity at the base of the brain. The hypothalamus influences the posterior pituitary through two hypothalamic nuclei: the **paraventricular nuclei** and the **supraoptic nuclei**. These nuclei make two hormones, oxytocin and vasopressin. **Oxytocin** causes contractions of a woman's uterus during childbirth. It also stimulates ejection of milk, or lactation, from her breasts during breast feeding. **Vasopressin** regulates the reabsorption of water by the kidneys, which helps to maintain blood pressure.

By an intricate blood vessel network called the **hypothalamopituitary portal system**, **releasing hormones** from the hypothalamus are transported to the anterior pituitary. These hormones then

The Case of H.M.

H.M. was a patient who underwent very radical brain surgery in 1953 at age 27. The surgery was designed to help control severe, debilitating epilepsy that had not responded to other forms of treatment. Large parts of his median temporal lobe in both hemispheres were removed, along with most of his left and right hippocampus. After the surgery, his intellect was not impaired and remained normal. However, he was left with severe amnesia called anterograde amnesia. He could still recall old memories but could no longer form new memories beginning on the day he emerged from surgery. He could remember new memories he formed for only a few minutes, then he forgot. His doctor had cared for him for many years, but each meeting was like the first, following the surgery. The type of memory that was affected is known as declarative memory, which consists of memories available to conscious awareness.

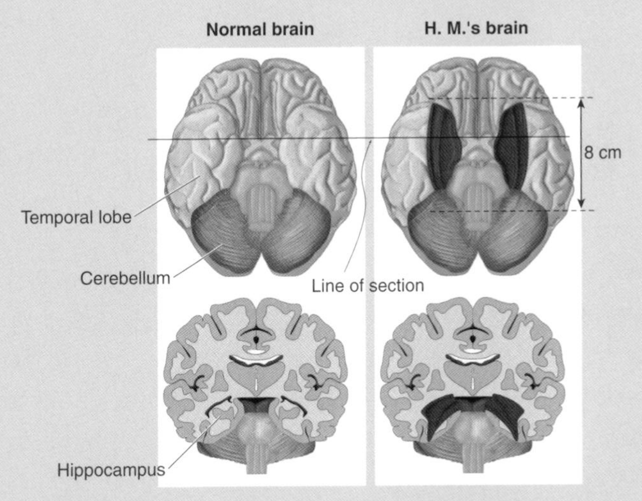

This figure shows a normal brain as compared to H.M.'s brain after surgery. H.M.'s surgeon performed a temporal lobe resection, in which he removed the hippocampus (red areas) and surrounding cortex in an effort to control his epileptic seizures.

trigger the release of a corresponding tropic anterior pituitary hormone that enters the blood and goes to a specific area of the body. For example, releasing hormone causes follicle-stimulating hormone (FSH) to travel to the ovaries and act on the ovarian tissue to produce follicles (see "The Case of H.M." box).

7 The Cerebral Cortex

The cerebral cortex is the outer covering of the cerebral hemispheres. It weighs about 600 grams, or 40% of the brain by weight. The cortex is essential to intellect, memory, consciousness, and learning, as well as processing sensory information. The cortex appears gray because it is mostly composed of cell bodies, dendrites, and unmyelinated axons. The structures below the cortex or gray matter are called subcortical and consist of white matter: these are the **basal ganglia** and the corpus callosum. The subcortical white matter contains large myelinated axons, which serve to connect parts of the cerebral cortex internally and with different parts of the brain.

An external examination of the brain reveals that the cerebrum has a highly convoluted surface. The large grooves across the cerebral hemispheres are called **fissures** and the smaller grooves are called **sulci** (singular: sulcus). A longitudinal fissure runs between the two cerebral hemispheres, separating them in the midline. There is also a left and right central fissure and a left and right lateral fissure. The raised ridges, or convolutions, between the adjacent fissures or sulci are called **gyri** (singular: gyrus) (Figure 7.1).

In humans and large mammals, the cerebral cortex became highly folded as it evolved, in order to accommodate its increased surface area. Most other vertebrates have brains with

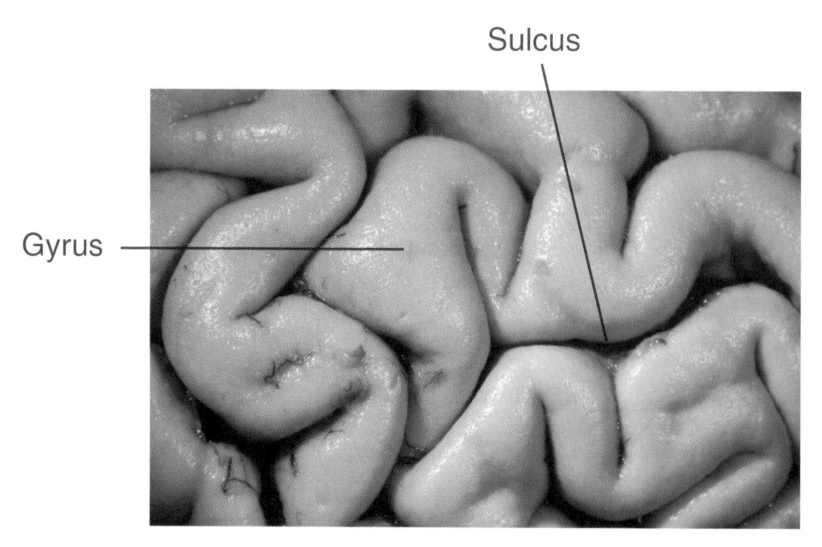

Figure 7.1 This photograph shows the surface of the brain, highlighting the sulci (foldings or troughs) and gyri (outgrowths or protrusions).

a smoother surface. A **coronal section** (Figure 7.2) of the human brain, or one cut parallel to the face, reveals that much of the cortex is actually hidden within the sulci and the fissures.

■ **Learn more about folds of the brain** Search the Internet for *brain fissures* and *sulci*.

Each cerebral hemisphere is divided into regions, or **lobes**, by the central and lateral fissures. The lobes describe the way the brain looks, and they help to identify the location of functional areas within the hemispheres, but the lobes themselves do not have specific functions (see "Cerebral Dominance" box).

The functional areas of the cerebrum are determined through clinical experimentation or through observation of people with brain damage.

The frontal lobes lie anterior to the central fissures. The most frontal part of the frontal lobes is called the "**prefrontal cortex.**" The **occipital lobes** lie at the posterior end of each hemisphere. The **pari-**

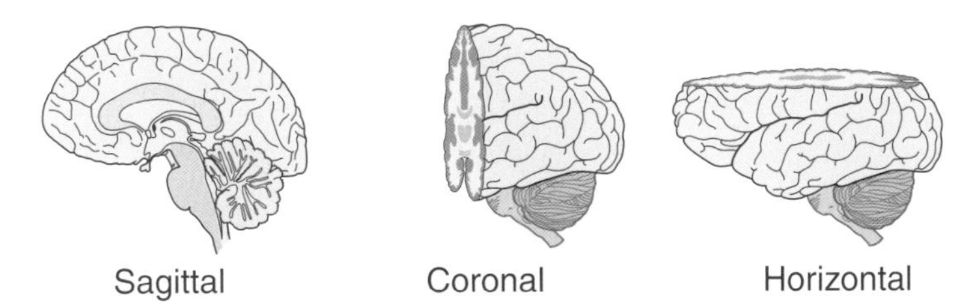

Figure 7.2. A coronal section of the brain (center) is a cut parallel to the face rather than a sagittal section, which is a cross section. A cross section or horizontal section is a cut parallel to the top of the brain.

etal lobes lie posterior to the central fissures and **superior** to the lateral fissures. The temporal lobes lie inferior to the lateral fissures and at the bottom part of each hemisphere (Figure 7.3).

In addition to the large central and lateral fissures, the cerebral hemispheres are also described according to the major cerebral gyri. The gyri lie adjacent to the fissures. Like the cerebral lobes, the gyri and the fissures do not represent functional units. The major gyri can be described in relation to the lobes. Each frontal lobe contains the precentral gyrus, the superior frontal gyrus, the middle frontal gyrus, and the inferior frontal gyrus. Each temporal lobe contains the superior temporal gyrus, the middle tempo-

Cerebral Dominance

Are you left-handed? Ninety percent of adults are right-handed. Their cerebral motor center on the *left* controls the right hand. In left-handed adults, the *right* hemisphere is dominant. Speech functions are found in the left hemisphere in most adults, and are not related to handedness. People who are ambidextrous (able to use either hand equally well) have 60% of their speech centers in the left hemisphere, 10% in the right, and 30% in both.

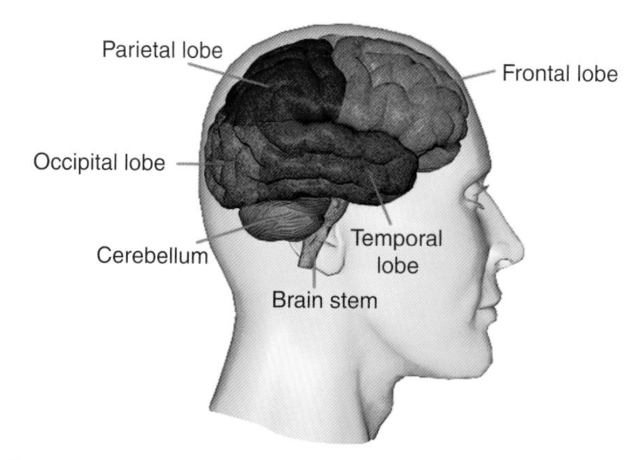

Figure 7.3 Each hemisphere of the brain is divided into four lobes by the central and lateral fissures. The lobes help to describe locations in the brain, but they do not have specific functions.

ral gyrus, and the inferior temporal gyrus. Each parietal lobe contains the postcentral gyrus and the angular gyrus.

With this anatomical outline of the cerebral hemispheres, we turn to a closer examination of the cortex itself. The primitive cortex, or **allocortex**, was the first type of cerebral cortex to evolve. It has three layers and is found in fish, reptiles, and amphibians.

A small part of the human cortex is composed of this primitive allocortex. The hippocampus, which is involved in memory and in navigation, is part of the allocortex. Most of the human cerebral cortex is **neocortex** (*neo* means "new"). The neocortex is structurally organized into six horizontal layers, numbered one to six. Layer six is the deepest, and layer one is the most superficial. Lobes and gyri describe areas of the cortex, and the cortical layers provide a differentiated cross section of these areas. Like the lobes and gyri, the cortical layers describe the structural appearance of the brain but do not have individual, discrete functions.

The cerebral cortex contains several types of neurons, but they fall into two main groups, characterized by the shape of the cell body and the course of the axon. These are **pyramidal cells** and **stellate cells.** Pyramidal cells have bodies shaped like pyramids, a long axon, and a central (or apical) dendrite projecting toward the surface of the cortex. Stellate cells have bodies shaped like stars, a short axon, and many dendrites. Stellate cells function within local circuits, synapsing with other neurons in the immediate vicinity. The pyramidal cells send signals to other structures in the CNS and also to other parts of the cortex (Figure 7.4).

The cerebral cortex can be subdivided into five functional areas that perform complex tasks: the primary and the secondary sensory areas, the primary and the secondary motor areas, and the **association areas.** The evidence for this classification is obtained from several clinical sources, including observations of patients with lesions producing seizures, patients who have cortical damage from injury or surgical procedures, or those who have had electrical stimulation applied to cortical areas.

The sensory areas are involved in receiving information from the sense organs (Figure 7.5). Sensory receptors in the body's sense organs are connected to the sensory areas via neural pathways. Each individual circuit, from a particular sense organ to the areas in the brain where its information is processed, is referred to as a sensory system. There are individual sensory systems for each of the five senses.

The **primary sensory areas** receive input from a single sensory system, usually via the thalamic relay nuclei. The sense of smell is the only sense that is not routed through the thalamus. Output from a primary sensory area goes to a **secondary sensory area** belonging to the same system. The secondary sensory areas receive input from a primary sensory area, plus other secondary sensory areas belonging to the same system. The primary sensory areas can be located as follows: taste in the post-central gyrus, smell on the ventral surface of the frontal lobe, sight in the

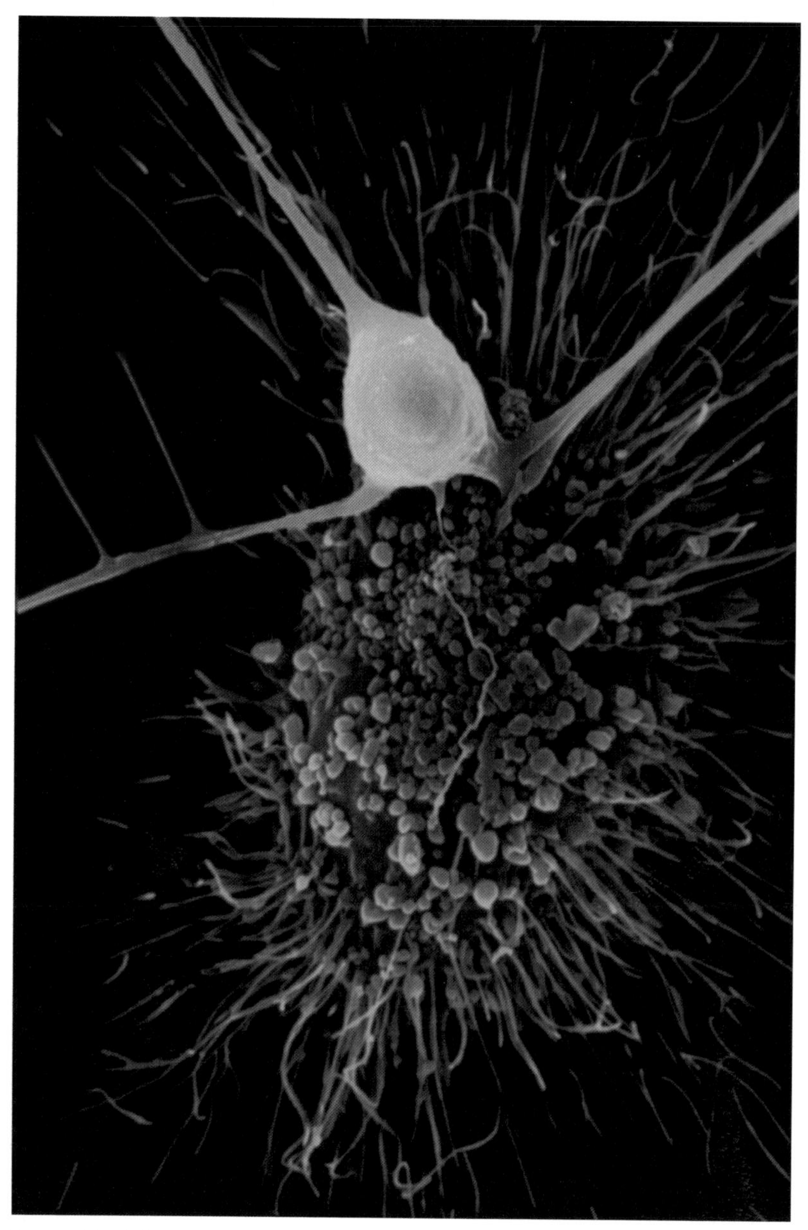

Figure 7.4 The neurons of the cerebral cortex are divided into two main groups, pyramidal cells and stellate cells, based on the shape of the cell body and the course of the axon. Pyramidal cells **(A)** have bodies shaped like pyramids, a long axon, and a central dendrite projecting toward the surface of the cortex.

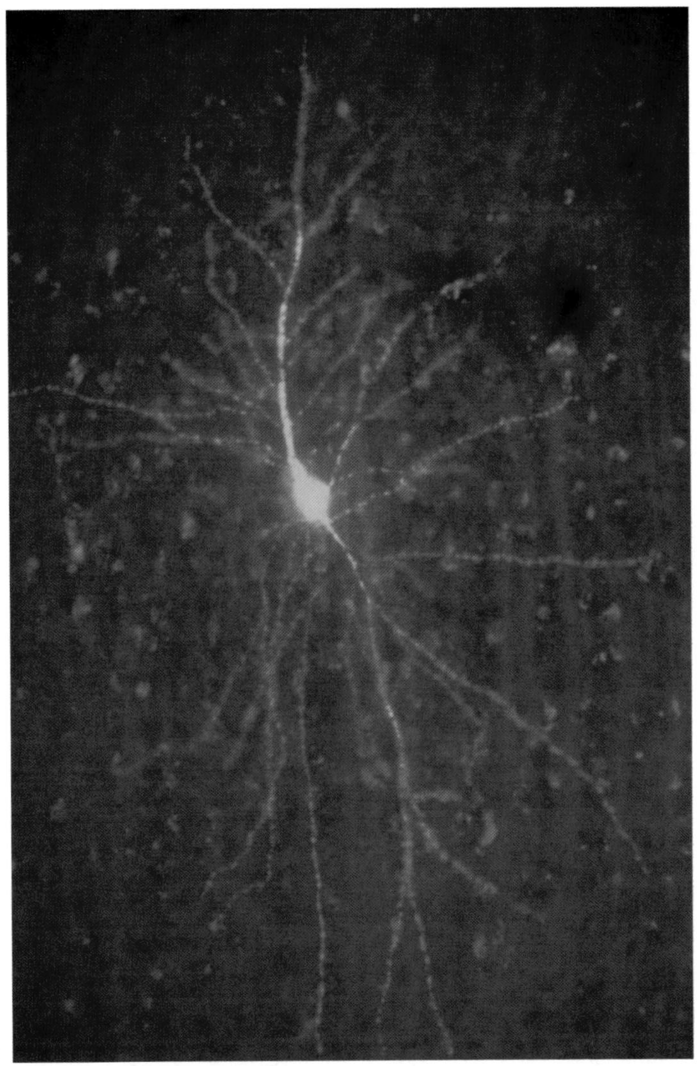

Figure 7.4 (B) Stellate cells have bodies shaped like stars, a short axon, and many dendrites.

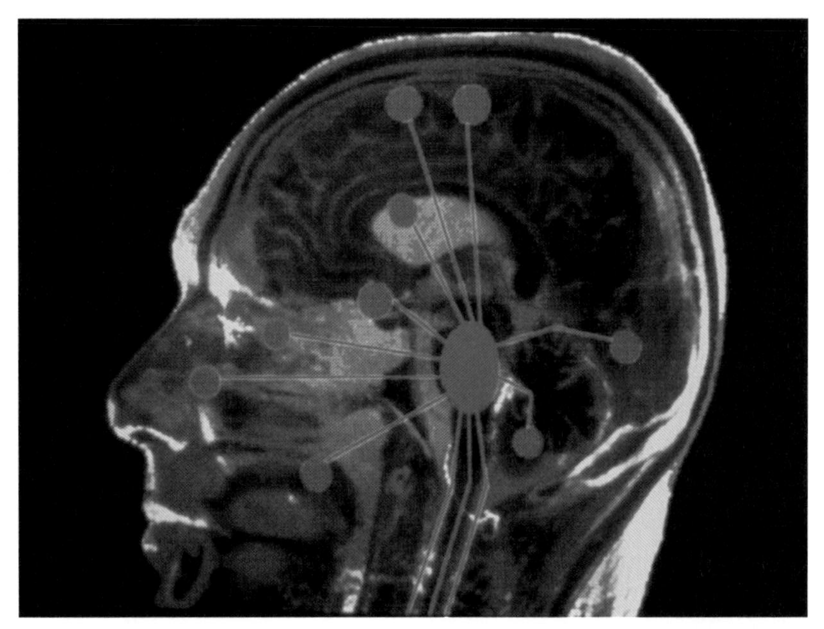

Figure 7.5 Three-dimensional NMR scan mapping sensory motor areas (green dots) in the brain.

occipital lobe, and hearing in the temporal lobe. The sensory areas that receive information pertaining to the sense of touch in each body part correspond with the motor areas that control the movement of each body part. They are located in the central and postcentral gyrus.

The motor areas control all voluntary movements of the body's muscles. They are located in the lateral area of the posterior prefrontal cortex. The **secondary motor areas** are responsible for selecting movement. The secondary motor areas receive input from association areas and send output to the **primary motor cortex**. The **primary motor areas** are responsible for executing movement. They receive input from their secondary motor areas and send output to subcortical and spinal motor circuits. Nerves connected to the spinal cord by synapses send commands directly to the muscles.

Association areas of the cerebral cortex receive input from more than one sensory system and send output to the secondary motor cortex. Association areas make up a large portion of the cortex. The first association area, consisting of the parietal, temporal, and occipital lobes, is involved in producing perceptions from sensory information. The second association area, in the **frontal lobe**, is involved in planning actions and movement. The third association area, in the limbic system, is involved in emotion and memory. The association areas perform very complex roles that are not well understood.

8 The Limbic System

The **limbic system** is important to the emotional life of humans and is especially involved in aggression, fear, and pleasure. It is also involved in the formation of memories. The limbic system has been called the emotional brain. It receives samples of sensory information, and its output influences endocrine, visceral motor, and somatic motor effectors. The limbic system is actually a midline circuit in the **temporal lobe**, which encircles the thalamus. It is about the size of a walnut. It includes the amygdala, the hippocampus, the cingulate cortex, the fornix, the septum, and the **mamillary bodies** (part of the hypothalamus).

The **amygdala** is directly involved in the emotions, especially fear. Anxiety, post-traumatic stress disorder, and phobias are related to damage to, or neurotransmitter imbalance within, the amygdala. The amygdala is also involved in the sense of smell. The **fornix** is the major tract of the limbic system. It releases a chemical that permits sexual stimulation. The exact role of the hippocampus is not fully known, but it is involved in learning and memory processing, sexual behavior, and emotions. The **hippocampus** is named after the sea horse (sea horse is *hippokampus* in Greek) because the shape of the hippocampus resembles this animal. It faces the **dentate gyrus**, and together, these are called the **hippocampal-dentate complex**. It

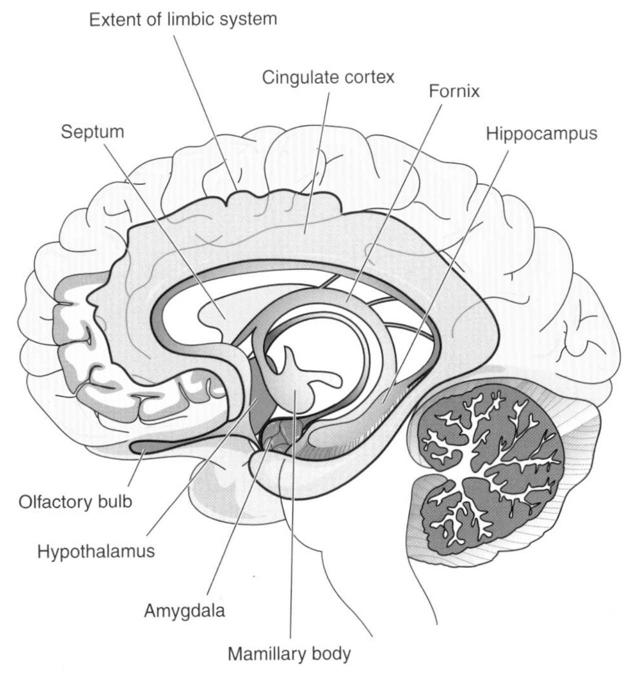

Figure 8.1 The limbic system is involved in the formation of emotions and memories. It includes the amygdala, the hippocampus, the cingulate cortex, the fornix, the septum, and the mamillary bodies.

has three layers of cells and is the oldest cortical portion of the limbic system. The afferent fibers in the hippocampus arrive from the para-hippocampal gyrus, the septal-hippocampal pathway, and from the contralateral hippocampus by way of the hippocampal commissures. The efferent fibers leave the hippocampus as the alveus and the fimbria (Figure 8.1).

The limbic system is involved in activities essential to individual survival—such as feeding, fighting, and fear—and also in activities related to species survival, such as mate selection, reproduction, and the care of infants. The limbic system is one of the oldest parts of the brain, which means that it appeared very early on the evolutionary scale. It is present in fish, amphibians, and reptiles as well as mammals. In humans, the limbic system is connected to the prefrontal cortex. This means that the least evolved brain system has a direct impact on the most evolved brain system. Some scientists speculate that this is why pleasure accompanies the solving of problems.

What is emotion? An emotion is a signal. Not surprisingly, emotions are related to essential needs such as food, reproduction, and survival. Emotions are motivational, in that we approach and repeat behaviors that bring us pleasure, such as eating sugar, but we avoid noxious stimuli that bring us pain, such as the stings of fire ants. Pleasure causes the desire to repeat, and pain causes fear, or the desire to avoid. Emotion is a response to a direct stimulus. How are feelings different from emotions? Feelings are complex interpretations of emotions, and they may be mixed with ambivalence. For example, a person may feel guilty about the impending death of a close family member suffering from terminal cancer. The guilt may involve a wish for the person's suffering to end, remorse for past conflicts with the person, regret about not spending more time with the person during his or her illness, and shame for having dismissed the first signs of the illness as having been "all in his or her head." An emotion also has characteristic, automatic manifestations, such as a dry mouth, moist palms, and an upset stomach during fear.

Aggressive monkeys who have ablation (surgical removal) of their bilateral anterior temporal lobes become docile and lose emotional responses such as anger and fear. This condition is called the **Klüver-Bucy syndrome.** Rhesus monkeys, who naturally fear snakes, will handle them after ablation of the bilateral anterior temporal lobe. Even if the snakes attack, the monkeys will go back and handle them again. Also, the rhesus monkeys' sexual behavior becomes significantly increased, and they will engage in increased masturbation, and heterosexual and homosexual

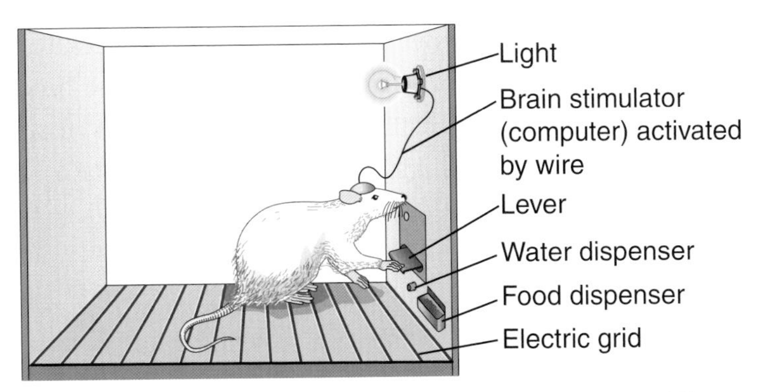

Figure 8.2 In 1954, the scientists James Olds and Peter Milner discovered that rats could be rewarded by stimulating specific areas in the animal's brains. The rats would push levers or navigate mazes in order to receive electrical stimulation to these brain areas.

behavior. The monkeys become unable to recognize objects using their eyes, and so will explore them with their mouths instead. Klüver-Bucy syndrome has also been observed in humans who suffer from herpes encephalitis or brain damage. These patients have visual recognition problems, put objects in their mouths, engage in inappropriate sexual behavior, lose normal fear and anger responses, and lose memory.

■ **Learn more about Klüver-Bucy syndrome** Search the Internet for *Klüver-Bucy.*

The limbic system contains pleasure centers. Stimulation of regions of the limbic system via implanted electrodes can identify where pleasure centers are located. Animals involved in these studies will push a lever repeatedly to cause electrical stimulation of their pleasure centers (Figure 8.2). The pleasant stimulation produces positive reinforcement for the animal to push the lever. Rats involved in these experiments will prefer stimulation of the pleasure center over food and will eventually die. Pleasure centers have been identified in the cingulate cortex, hippocampus, amygdala, hypothalamus, and the anterior nuclei of the thalamus. Pleasure centers are involved in sexual stimulation and in the "high" associated with certain street drugs.

MEMORY

Memory is a well-studied concept in psychology, and different types of memory have been described. Declarative memory, or those memories available to conscious awareness, is either semantic or episodic. Episodic memories relate to the self and concern events that occurred during a person's life. Semantic memories encompass knowledge of concepts and facts not related to a specific time or place. For example, you may be able to list all the states in the United States, but cannot recall exactly when you learned each of the states' names. An example of episodic memory might be "the first time I ever tried ice skating I was 7-years-old, and I remember how hard it was to keep my balance." Procedural memories typically involve motor skills. For example, throwing a ball, walking, and running are all accomplished with the aid of procedural memories. These memories are not consciously perceived, and they occur automatically.

The hippocampus is a major component of all memory systems (see "The Maturing Hippocampus" box). It is very important in short-term working memory. Short-term and long-term memories are not stored in the hippocampus, but long-term memories may be stored in the dentate and parahippocampal gyri.

Learning and memory are linked. Working memory is temporary. It may be that you only need to remember a phone number long enough to dial it on your cell phone, but you did not need to actually learn it. The content of working memory changes constantly in order to help you perform whatever task is at hand. The association areas are involved in working memory.

THE SENSES, SPATIAL MEMORY, AND THE HIPPOCAMPUS

Nobel prize winner Dr. Eric Kandel studied the very simple brains of the giant snail, called alpysia, to demonstrate how the brain uses sensory information to learn (Figure 8.3). He studied specific sensory and motor neurons that cause the snail to move its gill when it perceives negative stimuli in its tail. In performing this reflex, the same circuit of cells was activated over and over again. Over time, the synaptic connections in the neurons of the activated circuit were strengthened and became more

The Maturing Hippocampus

In a recent study, researchers compared two groups of healthy adolescents.[1] The younger group had participants aged 10–13 years, and the older group had participants aged 19–21 years. The researchers analyzed high-resolution three-dimensional magnetic resonance imaging (MRI) scans to examine developmental brain changes that occur during this period.

The study found that the volume of the hippocampus was significantly larger in male older adolescents than in male younger adolescents. Also, significantly less cerebral gray matter volume and significantly greater cerebral white matter volume were found in older adolescents compared to younger adolescents. These findings suggest that maturation of the human hippocampus continues during adolescence and may be most robust in males. These results may also have implications for the development of psychiatric disorders in young people. Further research is needed to better understand brain development.

[1] Suzuki, M., H. Hagino, S. Nohara, et al. "Male-Specific Volume Expansion of the Human Hippocampus During Adolescence" *Cerebral Cortex* 15.2 (2005): 187–93.

Figure 8.3 The simplicity of the alpysia's nervous system and its rich behavioral repertoire made it ideal for Dr. Eric Kandel's studies of the key molecular mechanisms of learning and memory.

efficient, and the snail became able to avoid the negative stimuli that had initially caused the reflex. The snail's learning was represented in the increased strength and efficiency of the synapse. In a paper published in 1970, Dr. Kandel said, "The capability for behavioral modification seems to be built directly into the neural architecture of the behavioral reflex." Later, he specified that the pre-synaptic membrane of its synapses was strengthened, resulting in an increased release of neurotransmitter. He also distinguished between alpysia's long-term memory and short-term memory and was able to recreate both types of memory in an experimental circuit consisting of only a single sensory neuron and a single motor neuron. Long-term synaptic changes required protein synthesis, but short-term changes did not. The protein synthesis required for long-term synaptic changes was caused by gene activity in the nucleus of the neuron.

The studies of alpysia focused on the simplest kind of memory (procedural memory). Dr. Kandel began studying mice to focus on explicit (voluntarily retrieved) memory. He and his colleagues discovered that, similarly to what had occurred with alpysia, a genetically activated protein synthesis occurred during the formation of long-term spatial memory in the mouse hippocampus. In addition, the stability of its long-term memory correlated with the degree to which the mouse was required to attend to its environment. Attention and protein synthesis work together to form a stable long-term spatial memory.[1]

In Alzheimer's disease, the hippocampus is the first part of the brain to be affected, and consequently, short-term memory problems and difficulty with spatial navigation are among the first symptoms of Alzheimer's disease.

---

1 Kandel, E. R. "Eric R. Kandel: Autobiography." The Nobel Foundation, 2005. http://nobelprize.org/medicine/laureates/2000/kandel-autobio.html.

9 The Basal Ganglia

The basal ganglia are two groups of cerebral nuclei, one in each hemisphere, that play a role in controlling voluntary movement and establishing body postures as well as in learning. Basal ganglia include the amygdala, the caudate, the putamen, and the globus pallidus (Figure 9.1). There are two complete sets of basal ganglia, one below each of the left and right cerebral hemispheres. The basal ganglia are located lateral and slightly anterior to the thalamus, and they are involved in transmitting information between the thalamus and the cerebral cortex. The basal ganglia can be compared to a crossing guard at an intersection, since they monitor, allow, or inhibit voluntary movement initiated by motor cortex regions.

The amygdala is an almond-shaped nucleus of the anterior temporal lobes. It is considered a part of *both* the basal ganglia motor system *and* the limbic system. The **caudate** nucleus arises from the amygdala in the anterior temporal lobe and resembles a tail. It extends **posteriorly** from the amygdala and almost completely encircles the other nuclei of the basal ganglia. The **putamen** is a nucleus of the basal ganglia, located in each hemisphere lateral to the **globus pallidus**. It is connected to the anterior end of the caudate by a series of fiber bridges. The putamen and the caudate together are called the **striatum** because they have a striped appearance. The striatum is

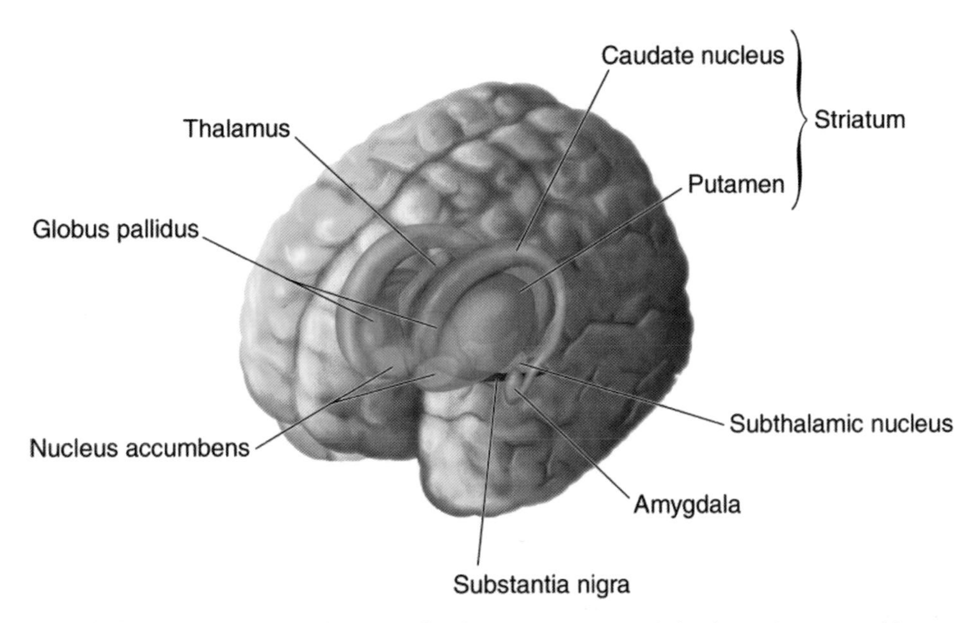

Figure 9.1 The basal ganglia control voluntary movement, body postures, and learning. Basal ganglia include the amygdala, the caudate, the putamen, and the globus pallidus. There are two complete sets of basal ganglia, one below each of the left and right cerebral hemispheres. The basal ganglia can be compared to a crossing guard at an intersection, since they monitor, allow, or inhibit voluntary movement initiated by motor cortex regions.

involved in the planning and modulation of movement pathways and also in executive function. The executive function manages and organizes other cognitive processes. In humans, the striatum is activated by stimuli that anticipate a reward and also by new and unfamiliar stimuli or challenges. Information from the cerebral cortex reaches the other basal ganglia via the striatum. The globus pallidus is also a nucleus of the basal ganglia. It is located in each hemisphere between the thalamus and the putamen. Globus pallidus means "pale globe."

It is still unknown exactly how the basal ganglia process information. Recent studies suggest that the basal ganglia are

Figure 9.2 This illustration shows patients suffering from chorea, a common symptom of HD. Patients with chorea exhibit jerky, involuntary movements of the shoulders, hips, and face. This condition is sometimes called St. Vitus's dance or St. Guy's dance.

highly interconnected with each other in ways that have yet to be determined. Disorders of movement that have been linked with the basal ganglia include Parkinson's disease (see "Parkinson's Disease" box), Tourette's syndrome, and Huntington's disease (HD). People with these disorders experience unwanted movements, such as jerking of the limbs or spasms of the facial muscles, as the normal balance of controls within the basal ganglia fails.

Huntington's disease is a genetic disease caused by an autosomal dominant mutation. The caudate nucleus degenerates, and cognition also becomes impaired. Symptoms of the disease do not usually appear until middle age, and then they progress until death about 10–20 years later. The disease is characterized by abnormal body movements, called chorea, which cause abnormal jerking of the limbs and facial muscles (Figure 9.2). No treatment for this disease is currently known.

Parkinson's Disease

Parkinson's disease is a movement disorder that can cause severe disability. It is a common neurological disorder that affects about 1% of the population over 60 years of age. It is the result of loss of dopamine-producing brain cells and depigmentation in the substantia nigra of the midbrain. There are four main symptoms of Parkinson's disease: slowness of movement; stiffness of the limbs and trunk; stooping; and a tremor of the hands, arms, legs, jaw, and face, which is worse at rest. Parkinson's disease is chronic and often progressive. The cause of Parkinson's disease is unknown in most cases, although strokes, brain infections, trauma, and neurotoxins may play a role in some cases. The *substantia nigra* is believed to be involved because, in advanced stages of Parkinson's disease, its neurons have degenerated. The substantia nigra is a dopaminergic midbrain nucleus, the neurons of which project to the striatum (the putamen and the caudate) via the nigrostriatal pathway. In chronic Parkinson's disease, almost none of the neurotransmitter dopamine is found in the substantia nigra and striatum.

The late Pope John Paul II, shown here, was affected by Parkinson's disease. A person with Parkinson's disease exhibits postural deficits and appears to be leaning forward in the direction of movement. When walking, the gait is slowed and the feet tend to shuffle as if weighted down.

■ **Learn more about disorders linked to the basal ganglia** Search the Internet for *Parkinson's disease, Huntington's disease,* or *Tourette's syndrome.*

There are two other disorders related to the basal ganglia that illustrate that the basal ganglia influence mental functioning such as learning and our state of mind. Both of the following disorders are mental illnesses in which motor functions play an important role.

Obsessive-compulsive disorder (OCD) is characterized by the unwanted need to constantly perform certain motor movements such as checking, inspecting, counting, collecting, cleaning and tidying, and repeating. People suffering from this disorder report that they cannot stop themselves from performing these often ritualistic and nonsensical movements. Neither can they stop themselves from thinking unwanted repetitive thoughts. The urge to perform compulsive acts may eventually interfere with a person's normal life so that it becomes severely disabling. Neuroimaging studies show increased activity in the putamen, caudate nucleus, orbital cortex, and cingulate gyrus when OCD symptoms are very strong. The basal ganglia are involved in helping to shift our attention from one activity to another, and people with OCD have great difficulty performing these shifts.

Attention deficit hyperactivity disorder (ADHD) can be considered the opposite of OCD. ADHD is a common disorder of children and is usually first identified by elementary school teachers who notice the child's abnormally poor attention span. ADHD children also have hyperactive motor movements and are impulsive compared to other members of their peer group. A child who cannot pay attention and is constantly distracted will go on to have trouble with learning and may eventually drop out of school in frustration. ADHD is a clinical diagnosis (performed through observation of patient behavior), but it would be helpful to also have a lab test or imaging test to support the diagnosis or to identify it in its very early stages. So far, neuroimaging studies using functional magnetic resonance imaging (fMRI) show reduced activity to prefrontal areas in children with ADHD compared to normal children, as well as reduced activity in parts of the basal ganglia.

10 Language and Thinking

Of all the species on the earth, the human species has the most sophisticated use of language and complex thought. The forebrain makes these functions possible.

The role of the forebrain in language is demonstrated most dramatically in a condition called **aphasia.** Damage to the left cortex can disrupt language and produce aphasia, or language dysfunction in which the ability to use or to comprehend words is impaired. Aphasia is not an abnormality of the structures involved in producing speech—the mouth and throat—although it may appear to be a "speech problem" to a casual observer. You may have met a person who became aphasic following a stroke—when a blood clot traveled into the brain and stopped or severely reduced blood flow to a part of the cortex.

There are several types of aphasia. **Receptive or sensory aphasia** is a deficit in the ability to symbolize, which is associated with lesions (injured brain tissue) of the **posterior temporoparietal region** in the **dominant hemisphere**. These lesions cause spoken language to be interpreted as meaningless or puzzling to comprehend. The patient may not be able to comprehend written language either. **Expressive or motor aphasia,** also attributed to lesions of the posterior temporoparietal region, prevents thoughts from being written or spoken in

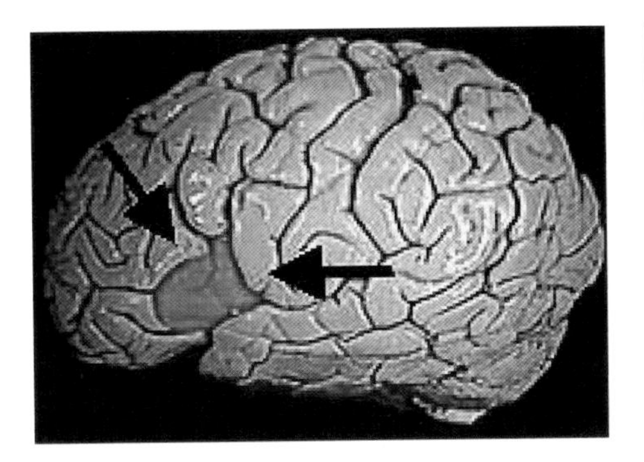

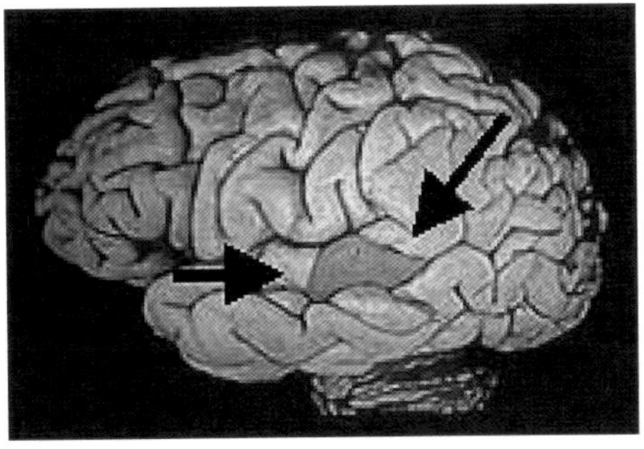

Figure 10.1 Broca's area (red) and Wernicke's area (blue) deal with language production and comprehension.

meaningful ways. In **nominal aphasia**, the patient is able to speak grammatically and fluently, but has trouble finding the right word for things (Figure 10.1). The patient can describe the thing he is trying to name but is unable to locate the correct noun that names the thing. In **averbia**, the patient has difficulty remembering only verbs. The common type of expressive aphasia is **Broca's aphasia.** The person with Broca's aphasia has great difficulty initiating and sustaining speech and is only able to emit fragments of what he is trying to express, often leaving out articles, pronouns, and conjunctions. The fragments he utters do have appreciable meaning. Patients who recover from Broca's aphasia say they knew what they were trying to say but were unable to get it out. **Wernicke's aphasia** is the common receptive aphasia. The person speaks with normal rate, inflection, and syntax, but many individual words in each sentence don't make any sense at all for their context, and he or she may utter words composed of nonsense syllables. Patients who recover from Wernicke's aphasia say that they were unable to comprehend their own speech or the speech of others. **Global aphasia** combines symptoms of both Wernicke's and Broca's aphasia.

THE WERNICKE-GESCHWIND MODEL

Let's examine one model for language, the **Wernicke-Geschwind model**, which identified functional areas involved in language based on observations of neurological patients with brain damage. The Wernicke-Geschwind model was proposed in 1965. Since that time, neuroscience has advanced rapidly, and imaging of the brain using fMRI (functional magnetic resonance imaging) and PET (positron emission tomography) has improved our understanding of the function of the different parts of the cerebral cortex.

fMRI enables a researcher to observe increased blood flow to various parts of the brain while the research subject is performing activities or responding to stimuli. The subject lies in a magnetic tube, and a series of scans are taken in succession. They can be compared, combined, or used to generate a three-dimensional image of the brain. With a PET scan (Figure 10.2), the subject is given an injection of a tracer substance. The tracer travels to the brain and is detected by the scanner. The location of the tracer provides information about the brain's metabolism.

The Wernicke-Geschwind model determined that seven areas of the left hemisphere are involved in language. The ground work was laid in the late 19th century by Carl Wernicke as he observed aphasia patients, and it was refined in the 1960s by Norman Geschwind, who combined linguistic and anatomical analysis. According to the Wernicke-Geschwind model, the **primary auditory cortex** controls hearing spoken words. The **primary visual cortex** controls seeing written words. The primary motor cortex controls use of the vocal apparatus (mouth, lips, tongue, and throat) to produce speech. **Wernicke's area** in the left temporal lobe, posterior to the primary auditory cortex, controls comprehension of spoken language. This is the part of the brain that is damaged in patients with Wernicke's aphasia. The **left angular**

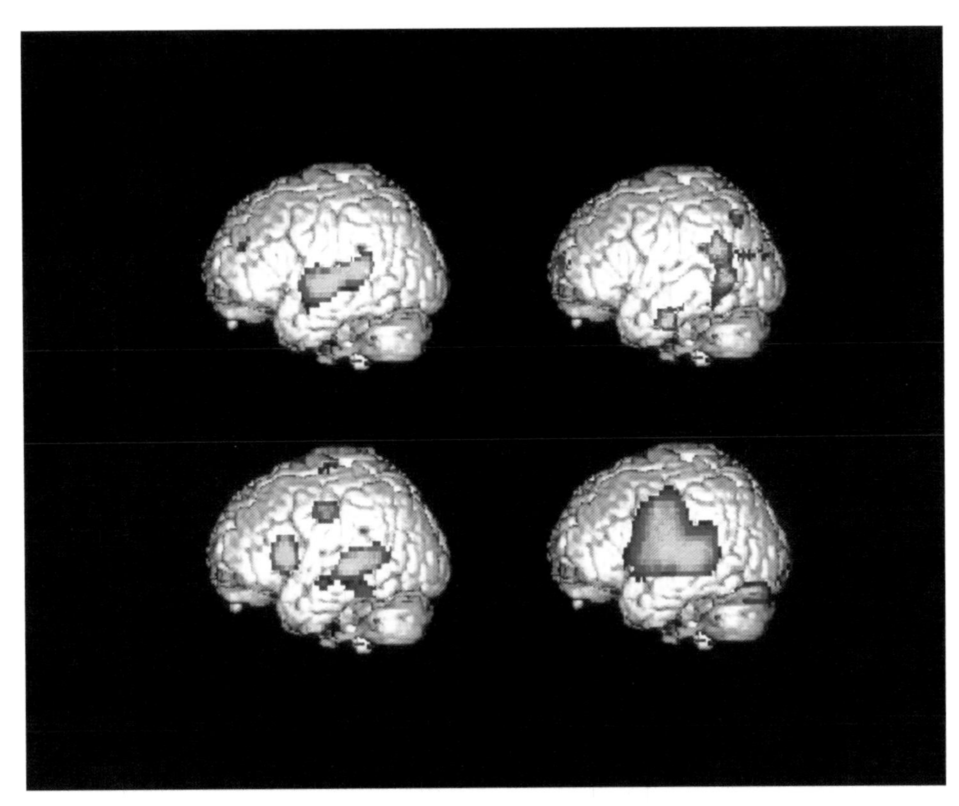

Figure 10.2. PET scans show brain activity when speaking or listening. At top left, monitoring imagined speech lights up the auditory cortex. At top right, working out the meaning of heard words activates other areas of the temporal lobe. At lower left, repeating words activates Wernicke's area for language comprehension (right), Broca's area for speech generation (left), and a motor region producing speech. At lower right, monitoring speech activates the auditory cortex.

gyrus in the parietal lobe performs the translation of written words into an auditory code. **Broca's area** in the left frontal lobe near the primary motor cortex is involved in the storage of programs for speech production and speech production itself. This is the part of the brain that is damaged in patients with Broca's aphasia. The **arcuate fasciculus**, a major tract that connects Broca's area with Wernicke's area, controls the activation of speech programs in Broca's area through the **Wernicke comprehension center.**

The Wernicke-Geschwind model has been challenged by

A Biological Link Between Brain Disorders

Schizophrenia is a severe brain disorder that usually begins in a patient's early 20s, but may also begin in childhood or adolescence. Persons with schizophrenia typically experience delusions (false, fixed beliefs) and hallucinations (false perceptual experiences) that are typically auditory. It is common for schizophrenics to hear voices telling them to do things. In addition, the normal executive functions of the brain become disorganized, and the person may experience apathy and difficulties with most of the activities of daily living. Schizophrenia is a *psychotic illness*, or an illness that affects thinking. Schizophrenia causes an enlargement of the fluid-filled ventricles of the brain (see illustration).

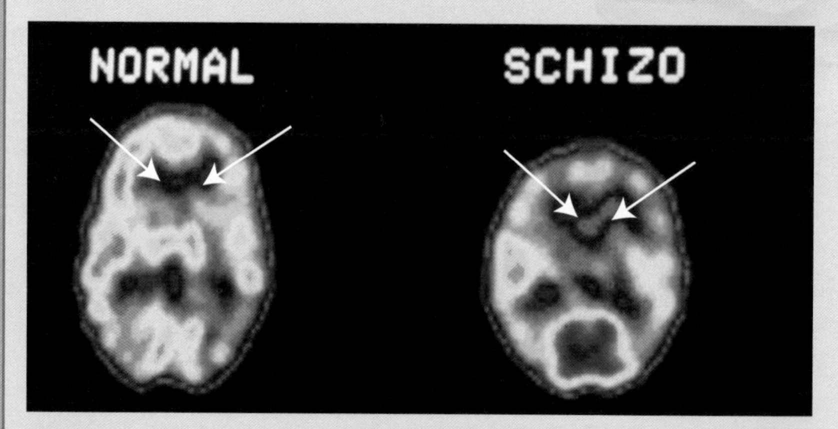

Bipolar disorder is also a brain disorder that affects young people, and it may become severely disabling. It is mostly a dysregulation of mood and energy level. People suffering from bipolar disorder may have episodes of depressed mood that alternate with manic episodes. During manic episodes, the person is of-studies using PET scans of subjects engaged in language tasks. According to the Wernicke-Geschwind model, all language activity occurs in the left cerebral hemisphere; but a study by Petersen, et al., showed that brain activity was bilateral (present in

ten very agitated and has a high energy level. About half of people who are manic also suffer from hallucinations.

Although psychiatrists traditionally regard schizophrenia and bipolar disorder as separate disorders, there may be a continuum of symptoms and a common biology underlying them. This common biology is evident in how the two disorders are treated. For example, some patients with schizophrenia are treated for depression, while some patients with bipolar disorder may receive antipsychotic medication to control their hallucinations.

Recently, Dr. Amy Arnsten, a researcher at Yale University in Connecticut, discovered a molecular trigger for *both* diseases, which may be activated by stress. Both schizophrenia and bipolar patients have elevated levels of an enzyme called protein Kinase C, or PKC, in the prefrontal cortex. High PKC activity may produce dysfunction of the prefrontal cortex, causing such symptoms as poor judgment, impulsivity, poor attention, and possibly hallucinations.

Arnsten discovered that high levels of stress can elevate PKC levels, which could explain why stressful events tend to trigger both diseases. If PKC could be blocked in the prefrontal cortex, then both diseases might be ameliorated. Dr. Arnsten notes that high PKC levels in the prefrontal cortex are also associated with lead poisoning. Patients with lead poisoning have symptoms of poor impulse control and poor attention span. Dr. Arnsten and other researchers point out that these findings not only point to possible links, triggers, and potential treatments for these disorders, they also help the public to understand that brain disorders have a chemical basis and are not the result of personal weakness.

both hemispheres) and that areas previously not believed to be involved in language were in fact activated.[1]

In the Petersen study, subjects who viewed nouns on a screen had bilateral activity in the primary visual cortex. Subjects who read the nouns aloud showed additional bilateral activity in the primary auditory cortex, the primary motor cortex, the **medial frontal cortex**, and the somatosensory cortex. Subjects who only viewed a blank screen showed little or no activity in the primary visual cortex.

In a study of thinking, or cognition, by Roland and Friberg, researchers asked subjects to think about three types of scenarios while the distribution of blood in the cerebral cortex was measured.[2] Radioactive substances were injected into the carotid artery supplying a particular cerebral hemisphere, and the distribution of radioactivity was measured electronically. Subjects were asked to count backward from 50 by threes, think about every second word in a jingle, and walk out their front door and turn left or right at each corner they encountered. All these tasks caused bilateral activity increases in the superior prefrontal cortex. Thinking about a route also increased activity bilaterally in the inferior prefrontal cortex, posterior parietal cortex, and inferior temporal cortex.

Like the Petersen study, the Roland and Friberg study shows that language functions involve both hemispheres of the brain. It also shows that other cognitive functions besides language involve both hemispheres. Although other studies have confirmed

---

1 Petersen, S. E., P. T. Fox, M. I. Posner, et al. "Positron Emission Tomographic Studies of the Cortical Anatomy of Single-Word Processing" *Nature*, 331 (1988): 585–589.

2 Roland, P. E. and L. Friberg. "Localization of Cortical Areas Activated by Thinking" *Journal of Neurophysiology*, 53 (1985):1219–1243.

the viability of the Wernicke-Geschwind model, these studies also suggest that ascriptions of certain types of functions to either the left hemisphere or the right hemisphere alone may require refinement.

The prefrontal cortex is typically associated with the center of human intelligence, but lesions in this area have negligible impact on intelligence tests. However, damage in specific areas of the cortex can produce other kinds of related deficits. The **dorsolateral prefrontal cortex** has a role in creative thinking and the ability to follow an action plan.

Lesions in the **orbitofrontal cortex** can produce personality change and inappropriate social behavior. Lesions in the medial prefrontal cortex may produce emotional blunting—for example, a patient may react not at all, or very minimally, to events that would produce strong emotion in a healthy person. (See "A Biological Link Between Brain Disorders" box.)

Glossary

Action potential Triggers exocytosis.

Allocortex The first type of cerebral cortex to evolve; has three layers; found in fish, reptiles, and amphibians.

Alpha-synuclein Type of protein that is the main component of the Lewy body.

Alzheimer's disease A disease affecting the basal forebrain in which neurons are disabled and the brain atrophies.

Amygdala One of the basal ganglia; considered a part of the basal ganglia motor system and the limbic system.

Anencephaly A neural tube defect (NTD) where the cerebral cortex does not form.

Anterior Toward the nose end in the nervous system of vertebrates.

Anterior commissure Connects the left and right temporal lobes.

Aphasia A language dysfunction usually resulting from brain damage.

Apolipoprotien E Referred to as apoE, a protein that helps carry cholesterol in the blood; linked to Alzheimer's disease.

Arachnoid membrane The middle membrane of the meninges; resembles a spider web.

Arcuate fasciculus Controls the activation of speech programs in Broca's area through the Wernicke comprehension center, according to the Wernicke-Geschwind model.

Association area One of the five functional areas of the cortex; performs very complex roles that are not well understood.

Attention deficit hyperactivity disorder (ADHD) Disorder characterized by abnormally poor attention span, hyperactive motor movements, and impulsivity; shows decreased activity in the basal ganglia.

Averbia A specific form of aphasia in which the patient has difficulty remembering only verbs.

Axon A long, single extension from the cell body which ends in axon branches; conducts signals away from the soma and transmits them to other cells.

Axon hillock A cone-shaped junction between the cell body and the axon; the neuron's trigger zone.

Basal ganglia Two groups of cerebral nuclei, one in each hemisphere, which play a role in controlling movement; they lie beneath the cortex.

Beta amyloid A toxic molecule formed from normal protein in the brain of an Alzheimer's patient.

Bipolar disorder A dysregulation of mood and energy level; characterized by episodes of depressed mood that alternate with manic episodes.

Broca's aphasia An aphasia in which the patient has difficulty initiating and sustaining speech and is only able to emit fragments of what he is trying to express.

Broca's area Involved in the storage of programs for speech production and speech production itself, according to the Wernicke-Geschwind model.

Buttons Terminal areas of axon branches, also called bouton terminals; the axon's buttons form the presynaptic side of the synapse.

Caudate One of the nuclei of the basal ganglia.

Cell body The metabolic center of the neuron, also called the soma.

Central nervous system (CNS) The part of the nervous system consisting of the brain and the spinal cord.

Cerebral cortex The surface layer of gray matter that covers the cerebral hemispheres; essential to intellect, memory, consciousness, voluntary actions, and learning.

Cerebral hemispheres Together, form the telencephalon.

Cerebrospinal fluid (CSF) Fills ventricles and spinal cord; acts like a support and cushion.

Cerebrum The two cerebral hemispheres.

Commissures Tracts which run through the longitudinal fissure, connecting the cerebral hemispheres.

Coronal section Section cut through the brain parallel to the face.

Corpus callosum The largest commissure; contains about 200 million axons.

Cranial nerves 12 pairs of nerves that exit directly from the brain; extremely important for functions such as smell, vision, eye movement, taste, salivation, swallowing, and hearing.

Cytoplasm The protoplasm outside the nucleus of a cell.

Dementia A progressive decline in mental functioning that includes severe memory impairment.

Dendrites Short, branch-like extensions of the cell body which receive signals from other neurons; the dendrites form the postsynaptic side of the synapse.

Dentate gyrus One of the two interlocking gyri that composes the hippocampus.

Diencephalon Posterior forebrain swelling in the neural tube; appears in the fifth week of prenatal growth; develops into the hypothalamus and thalamus; part of the brain stem.

Dominant hemisphere Cerebral hemisphere located on the opposite side of the body as the hand uses to write.

Dopamine A neurotransmitter, the lack of which is instrumental in Parkinson's disease.

Dorsal On the back or top of the head in the human nervous system.

Dorsolateral prefrontal cortex Plays a role in creative thinking and the ability to follow an action plan.

Dura mater The outermost and toughest of the meninges; contains strong connective tissue and lies next to the skull.

Excitatory post-synaptic potential (EPSP) A charge which excites the post-synaptic membrane.

Exocytosis The process of neurotransmitter release.

Expressive of motor aphasia Aphasia which prevents thoughts from being written or spoken in meaningful ways.

Exuberant period Time of accelerated formation of synaptic connections, between birth and two years old.

Fissures Large grooves across the cerebral hemispheres.

Folic acid Vitamin which aids in prevention of neural tube defects (NTDs).

Forebrain The part of the brain that controls thought, memory, and language.

Fornix The major tract of the limbic system.

Frontal lobes Lobes anterior to the central fissures of the cerebrum.

Gastrulation The process through which a developing embryo undergoes an elaborate series of infoldings that generate three layers of cells called the ectoderm, mesoderm, and endoderm.

Global aphasia Combines symptoms of both Wernicke's and Broca's aphasia.

Globus pallidus A nucleus of the basal ganglia.

Gray matter Composed of cell bodies, dendrites, and unmyelinated axons.

Gyri Raised ridges, or convolutions, between adjacent fissures or sulci.

Hippocampal-dentate complex The oldest cortical portion of the limbic system, it has three layers of cells.

Hippocampus An important part of the limbic system; a curved elongated ridge which lies along the medial edge of the temporal cortex; involved in memory and navigation.

Homeostasis Maintenance of a steady state in the body.

Hypertension High blood pressure.

Hypothalamopituitary portal system Intricate blood vessel network that transports releasing hormones from the hypothalamus to the anterior pituitary gland.

Hypothalamus Located adjacent to the third ventricle, between the cerebrum and the brain stem; regulates the posterior pituitary gland and important in maintaining homeostasis; together with the thalamus, forms the diencephalon.

Inferior Below.

Inhibitory post-synaptic potential (IPSP) A charge which inhibits the post-synaptic membrane.

Klüver-Bucy syndrome Syndrome discovered in monkeys that have had their bilateral anterior temporal lobes removed; characterized by docility, loss of emotional response, and increase in sexual behavior.

Left angular gyrus Performs the translation of written words into an auditory code, according to the Wernicke-Geschwind model.

Lewy body type dementia Dementia characterized by formation of Lewy bodies.

Limbic system Important to the emotional life of humans; encircles the thalamus; includes the amygdala, the hippocampus, the cingulate cortex, the fornix, the septum, and the mamillary bodies.

Lobes Regions of the cerebral hemispheres; help to identify the brain's functional areas but do not themselves have specific functions.

Macroglia Neuroglia made up of astrocytes.

Mamillary bodies In the hypothalamus; play an important role in emotional regulation.

Massa intermedia Commissure in the middle of the third ventricle; connects the left and right lobes of the diencephalon; also a thalamic nucleus.

Medial frontal cortex Plays a roll in emotional response.

Meninges Three membranes which cover and protect the CNS.

Mesencephalon Midbrain swelling in the neural tube; appears in the fourth week of prenatal growth.

Metencephalon Anterior hindbrain swelling in the neural tube; appears in the fifth week of prenatal growth.

Microglia The smallest neuroglia, they scavenge and ingest foreign matter, becoming distended when they do so.

Microtubules Thin tubes of protein that help to provide transportation routes for molecules within the neuron.

Midline The axis of symmetry of the brain, between the left and right hemispheres.

Mitochondria Small, intracellular part of the cell involved in the production of energy for the cell.

Multi-infarct dementia Dementia caused by a series of strokes.

Myelencephalon Posterior hindbrain swelling in the neural tube; appears in the fifth week of prenatal growth.

Myelin Dense, white, fatty substance; wraps axons and promotes transmission of action potentials.

Neocortex Most of the human cerebral cortex; has six layers.

Neural crest Part of the embryo which will become the peripheral nervous system.

Neural plate Patch of special cells on the surface of the embryo which will become the neural tube and the neural crest.

Neural tube Part of the embryo which will become the central nervous system.

Neural tube defect Defects of the brain and spinal cord resulting from incomplete closing of the neural tube in an embryo.

Neurofilaments Long, fine threads that provide a supportive matrix.

Neurogenesis The production of new neurons.

Neuroglial cells One of the two main types of brain cells; perform a supportive or protective function in the nervous system.

Neurologist A medical doctor who specializes in diagnosing and treating diseases of the brain and nervous system.

Neurons Specialized cells that receive, conduct, and transmit electrochemical signals; one of the two main types of brain cells.

Neurotransmitters Molecules that are released at the synapses; travel from pre-synaptic to post-synaptic membrane; can either help excite or inhibit other neurons; active transmitters include serotonin, dopamine, acetylcholine, and norepinephrine.

Neurulation Process in the development of an embryo during which the neural tube is formed.

Nodes of Ranvier Gaps between areas of myelin on the axon; action potentials jump from node to node.

Nominal aphasia Aphasia in which the patient is able to speak grammatically and fluently, but has trouble finding the right word for things.

Obsessive-compulsive disorder (OCD) Disorder characterized by the unwanted need to perform motor movements and to repeat thoughts; shows an increase in activity in the basal ganglia.

Occipital lobes Lobes that lie at the posterior end of each hemisphere of the cerebrum.

Oligodendrocytes Glial cells that myelinate central nervous system axons.

Orbitofrontal cortex Plays a role in personality and social behavior.

Oxytocin Hormone made in the hypothalamus; causes contractions of the uterus and stimulates lactation.

Paraventricular nuclei In the hypothalamus; make oxytocin and vasopressin.

Parietal lobes Lobes that lie posterior to the central fissures and superior to the lateral fissures of the cerebrum; contain the postcentral gyrus and the angular gyrus.

Parkinson's disease A movement disorder which can cause severe disability.

Peripheral nervous system (PNS) Nerves outside of the central nervous system.

Pia mater Bottommost and most delicate of the meninges; adheres to and covers the brain and spinal cord.

Plaques Clumps of beta amyloid.

Plasticity Changes in the connections between neurons as a result of experience.

Posterior Located toward the tail end in the nervous system of vertebrates.

Posterior temporoparietal region Region in which lesions produce aphasia.

Post-synaptic membrane The part of the cell membrane where the release of a neurotransmitter from the synapse creates either an excitatory post-synaptic potential (EPSP) or inhibitory post-synaptic potential (IPSP).

Preoptic area In the hypothalamus; plays a role in sexual behavior.

Prefrontal cortex Typically associated with the center of human intelligence; the "executive" part of the brain.

Pre-synaptic membrane Location on the button where synaptic vesicles release packets of neurotransmitter molecules.

Primary auditory cortex Controls the hearing of spoken words, according to the Wernicke-Geschwind model.

Primary motor area One of the five functional areas of the cortex; controls movement of certain body parts.

Primary motor cortex Controls use of the vocal apparatus to produce speech, according to the Wernicke-Geschwind model.

Primary sensory area One of the five functional areas of the cortex; receives sensory information from the thalamus.

Primary visual cortex Controls the seeing of written words, according to the Wernicke-Geschwind model.

Prosencephalon Forebrain swelling in the neural tube; appears in the fourth week of prenatal growth.

Pruning Process of eliminating unused synapses, increasing efficiency and processing speed in the brain; occurs during childhood and adolescence.

Psychiatrist A medical doctor who specializes in treating both physical and mental disorders of the brain.

Putamen A nucleus of the basal ganglia.

Psychotic illness An illness that affects thinking.

Pyramidal cells One of the two main types of cells in the cerebral cortex; have bodies shaped like pyramids, a long axon, and a central dendrite projecting toward the surface of the cortex.

Receptive or sensory aphasia Aphasia in which spoken (and sometimes written) language seems to be interpreted as meaningless or puzzling to comprehend.

Releasing hormones Made in the hypothalamus; trigger the release of corresponding hormones in the anterior pituitary gland.

Rhombencephalon Hindbrain swelling in the neural tube; appears in the fourth week of prenatal growth.

Schizophrenia A severe brain disorder that usually begins in a patient's early 20s, but may also begin in childhood or adolescence. Persons with schizophrenia experience delusions and hallucinations that are typically auditory.

Schwann cells Glial cells that myelinate peripheral nervous system axons.

Secondary motor area One of the five functional areas of the cortex; important in speech.

Secondary sensory area One of the five functional areas of the cortex.

Soma The metabolic center of the neuron, also called the cell body.

Spina bifida A neural tube defect (NTD) in which part of the spinal cord develops outside of the spine.

Stellate cells One of the two main types of cells in the cerebral cortex; have bodies shaped like stars, a short axon, and many dendrites.

Stem cells Fetal tissue created at the very earliest stages of embryonic life that has the potential to develop into any type of cell in the body.

Striatum The putamen and the caudate together, involved in the planning and modulation of movement pathways, and also in executive function.

Strokes A sudden loss of brain function caused by a blockage or rupture of a blood vessel to the brain, usually happening suddenly, allowing rapid changes in a person's functioning.

Subarachnoid space Between the arachnoid membrane and the pia mater; filled with cerebrospinal fluid.

Substantia nigra A dopaminergic midbrain nucleus; in Parkinson's disease, loses dopamine-producing cells.

Sulci Small grooves across the cerebral hemispheres.

Superior Above.

Suprachiasmatic nuclei In the hypothalamus; involved in circadian rhythms.

Supraoptic nuclei In the hypothalamus; make oxytocin and vasopressin.

Synapse Small gap between the buttons of one neuron and the receptive membrane of another.

Synaptic cleft The space between the pre-synaptic and post-synaptic sides.

Synaptic vesicles Store neurotransmitter molecules until they are ready for release.

Tau A type of protein which helps support neurons; becomes tangled in the brain of an Alzheimer's patient.

Telencephalon Anterior forebrain swelling in the neural tube; appears in the fifth week of prenatal growth; develops into the two cerebral hemispheres; is the largest division of the human brain.

Temporal lobes Lobes that lie inferior to the lateral fissures and at the bottom part of each hemisphere of the cerebrum; each temporal lobe has a superior temporal gyrus, a middle temporal gyrus, and the inferior temporal gyrus.

Thalamus An integrator and hub involved in almost all the activities of the forebrain; relays incoming sensory pathways to the cerebral cortex; mostly comprised of gray matter, but internal layers containing myelinated axons give it a striped appearance; together with the hypothalamus, forms the diencephalon.

Thrombospondins Proteins that signal synapse formation.

Transthyretin Brain protein that stops the progression of Alzheimer's disease in human brain tissue.

Vasopressin Hormone made in the hypothalamus; regulates the reabsorption of water by the kidneys.

Ventricles Four spaces in the cerebrum filled with cerebrospinal fluid (CSF).

Ventromedial nuclei In the hypothalamus; help to regulate the conversion of blood glucose to body fat.

Wernicke-Geschwind model A model for how the brain controls language processes, now believed to be inaccurate.

Wernicke's aphasia The common receptive aphasia, in which the person speaks with normal rate, inflection, and syntax, but many individual words in each sentence don't make sense for their context, and sometimes utter nonsense syllables.

Wernicke's area Controls comprehension of spoken language, according to the Wernicke-Geschwind model.

Wernicke comprehension center Medium through which the arcuate fasciculus controls the activation of speech programs.

White matter Areas of the nervous system that have many myelinated axons.

Bibliography

Christopherson, K., E. M. Ullian, C.C.A. Stokes, et al. "Thrombospondins Are Astrocyte-Secreted Proteins that Promote CNS Synaptogenesis." *Cell* 120(33)(2005): 421–433.

Dubin, M. W. *How the Brain Works.* Malden, MA: Blackwell Science Inc., 2002.

Johnson, M. H. "Functional Brain Development in Humans." *Neuroscience* 2 (2001).

Kandel, E. R. "Eric R. Kandel: Autobiography." *The Nobel Foundation* (2005). http://nobelprize.org/medicine/laureates/2000/kandel-autobio.html.

Langenscheidt's Pocket Merriam-Webster Medical Dictionary. New York, NY: Langenscheidt Publishing Group, 2002.

Noback, C. R. and R. J. Demarest. *The Nervous System.* New York, NY: McGraw-Hill, 1977.

Petersen, S. E., P. T. Fox, M. I. Posner, et al. "Positron emission tomographic studies of the cortical anatomy of single-word processing." *Nature,* 331 (1988): 585–589.

Pinel, J.P.J. *A Colorful Introduction to the Anatomy of the Human Brain.* Boston, MA: Allyn & Bacon, 1998.

Roland, P. E. and L. Friberg. "Localization of cortical areas activated by thinking." *Journal of Neurophysiology,* 53 (1985):1219–1243.

Suzuki, M., H. Hagino, S. Nohara, et al. "Male-Specific Volume Expansion of the Human Hippocampus During Adolescence." *Cerebral Cortex* 15.2 (2005): 187–193.

"What's in Your Mind?" *National Geographic,* March 2005.

Further Reading

Brynie, F. H. *101 Questions Your Brain Has Asked About Itself but Couldn't Answer . . . Until Now.* Brookfield, CT: The Millbrook Press, 1998.

Conlan, R., ed. *States of Mind: New Discoveries About How Our Brains Make Us Who We Are.* New York, NY: Dana Press, 1999.

Dubin, M. W. *How the Brain Works.* Malden, MA: Blackwell Science Inc., 2002.

Greenfield, S. A. *The Human Brain: A Guided Tour.* Science Masters Series, New York, NY: Basic Books, 1997.

Spector, M. "Rethinking the Brain: How the Songs of Canaries Upset a Fundamental Principle of Science." *The New Yorker,* July 15, 2001, 42–53.

Websites

The National Institute of Mental Health
www.nimh.gov

The National Institute of Neurological Disorders and Stroke
www.ninds.nih.gov

The National Center for Infants, Toddlers, and Families
www.zerotothree.org

Nature magazine
www.nature.com, subject area "neuroscience"

Index

About the Author

Dr. Elizabeth Tully is a National Institute of Mental Health post-doctoral research fellow at Yale University in New Haven, Connecticut. She received a Bachelor's degree in Psychology and a Master's degree in Biology from Wayne State University in Detroit, and an M.D. from the University of Turin, Italy, Faculty of Medicine and Surgery, in 1986. She also completed a psychiatric residency fellowship in child and adolescent psychiatry at the University of Utah in 1991. She is a contributing author to *Textbook of Schizophrenia*, edited by J. A. Lieberman, T. S. Troup, and D. A. Perkins, forthcoming in 2005 from American Psychiatric Publishing. Her special interest is the early onset of psychotic disorders, especially schizophrenia.

Picture Credits

page:

3: ©Peter Lamb
5: ©Roger Ressmeyer / CORBIS
7: Simon Lovestone, Institute of Psychiatry, London
12: ©Peter Lamb
15: Greg Gambino / 20•64 Design
17: Science VU / Visuals Unlimited
22: Greg Gambino / 20•64 Design
23: Greg Gambino / 20•64 Design
31: James Prince / Photo Researchers, Inc.
33: Greg Gambino / 20•64 Design
34: Zephyr / Photo Researchers, Inc.
36: Scott Camazine & Sue Trainor / Photo Researchers, Inc.
39: Scott Camazine & Sue Trainor / Photo Researchers, Inc.
41: ©Peter Lamb
44: Geoff Tompkinson / Photo Researchers, Inc.
45: Greg Gambino / 20•64 Design
46: Used with permission by TBI Resource Guide, www.neuroskills.com.
48: Dennis Kunkel / Phototake
49: Dennis Kunkel / Phototake
50: Collection CNRI / Phototake
53: ©Peter Lamb
55: Greg Gambino / 20•64 Design
58: Robert Barber / Visuals Unlimited
61: ©Peter Lamb
62: J. L. Charmet / Photo Researchers, Inc.
63: Ponkawonka.com
67: Neuroscience for Kids
69: WDCN / Univ. College London / Photo Researchers, inc.
70: John D. Cunningham / Visuals Unlimited